FIBER OPTIC ESSENTIALS

FIBER OPTIC ESSENTIALS

Casimer M. DeCusatis
Distinguished Engineer, IBM Corporation, Poughkeepsie, New York

and

Carolyn J. Sher DeCusatis
Consultant

AMSTERDAM • BOSTON • HEIDELBERG • LONDON
NEW YORK • OXFORD • PARIS • SAN DIEGO
SAN FRANCISCO • SINGAPORE • SYDNEY • TOKYO

ELSEVIER

Academic Press is an imprint of Elsevier

Academic Press is an imprint of Elsevier
30 Corporate Drive, Suite 400, Burlington, MA 01803, USA
525 B Street, Suite 1900, San Diego, California 92101-4495, USA
84 Theobald's Road, London WC1X 8RR, UK

This book is printed on acid-free paper. ♾

Library of Congress Cataloging-in-Publication Data
Application Submitted

British Library Cataloguing in Publication Data
A catalogue record for this book is available from the British Library

ISBN 13: 978-0-12-208431-7
ISBN 10: 0-12-208431-4

For information on all Elsevier Academic Press publications
visit our Web site at www.books.elsevier.com

Printed in the United States of America
05   06   07   08   09   10   9   8   7   6   5   4   3   2   1

Contents

Preface

The development of optical fiber technology for communication networks, medical applications, and other areas represents a unique confluence of the physics, electronics, and mechanical engineering disciplines. Its history extends back to the earliest uses of light and mirrors by ancient civilizations, including many false starts (the so-called "lost generation" described in Jeff Hect's book, *City of Light*), under-appreciated discoveries, and success stories. Today, the amount of optical fiber installed worldwide is equivalent to over 70 round trips to the moon, and this technology has matured to serve as the backbone of the Internet, the global telecommunications infrastructure, and the basis of exciting new developments in illumination, imaging, sensing, and control that were scarcely imagined when the laser was first developed in the early 1960s.

As this field has grown, many new practitioners with an established background in physics or electrical engineering have begun to work in some branch of the optical fiber field. Service technicians accustomed to working with copper cables have been asked to retrain themselves and understand the basics of optical cable installation. There are a growing number of non-traditional workers, including technical marketers, patent lawyers, and others who are attempting to transition into this area. And of course, many new students have begun to study this field as well. Understanding the fundamentals of fiber optics is essential for both the application designer and the basic researcher, and there are many good comprehensive references available, some of which extend to multiple volumes with thousands of pages. Clearly, entire books can and have been written on the topics of lasers, fiber manufacturing, and related areas. However, in working with both students and professionals who have recently entered this field, we have found a need for a much simpler introduction to the field. In this book, we have attempted to provide an overview of the field suitable for those with some technical or practical background that would allow a more rapid, accessible grasp of the material. We began with the assumption that chapters should be as short as possible, with minimal use of equations and no derivations. Condensing an entire book's worth of material into such a chapter is not an easy task, and we have needed to use a good deal of judgment in determining which topics to include and which to leave out. In this, we have tried to let practical applications rather than theoretical considerations be our guide. For example, we make only a passing reference to Maxwell's equations and do not use them to derive any results. This is not because we fail to appreciate the importance and elegance of these expressions; on the contrary, we would not be able to do them justice in the limited space allowed, and instead have left this topic open for readers who may wish to pursue it. Likewise, we offer only a brief nod to semiconductor materials engineering before describing its application to lasers and light emitting diodes (LEDs). The interested reader may refer to the many references provided on these and other topics, while bearing in mind that the guiding purpose of this book is to provide a high level overview for the general practitioner in this field.

We have also included some unique features which are not commonly found in handbooks of this type. Many prospective readers have asked us for a guide to understanding the bewildering number of acronyms and jargon in this field; accordingly, we provide an extensive list of definitions in the appendix which will hopefully make it easier for others to read in more detail about this field. We have provided a brief timeline of significant developments in the field, extending the excellent work done by others into the present day. Similarly, there is a chapter on basic facts about fiber optic technology which will hopefully be both interesting and informative to the casual reader, and offer some perspective on how this technology is applied to affect our daily lives. Along the same lines, we have included chapters on medical and other applications of fiber optics, something not often found in fiber handbooks. Overall, we hope this book strikes a balance between technical detail and reduction to engineering practice that will make it a useful source for many people in this area.

An undertaking such as this book would not be possible without a supportive staff at our publisher and the understanding of our families to whom we extend our deepest gratitude. This book is dedicated to our daughters, Anne and Rebecca, without whose inspiration and delight in math and science it would not have been possible. We also gratefully acknowledge the support of Dr. and Mrs. Lawrence Sher, Mrs. Helen DeCusatis, and the memory of Mr. Casimer DeCusatis, Sr.

Casmier M. DeCusatis
Carolyn J. Sher DeCusatis
Poughkeepsie, New York, June 2005

Chapter 1 | Fiber, Cables, and Connectors

Every fiber optic system has three basic components—a **source**, a **fiber**, and a **receiver**. Each of these components will be discussed in more detail in the following chapters; for now, we provide a brief overview of each one and describe how they work together.

All types of fiber optic systems require a light source; for applications such as medical imaging or architectural lighting, this source can be any type of conventional light bulb. The optical fiber serves as a light guide in this case; its purpose is simply to convey light from the source to a desired destination. In optical communications systems, the source of light used is called a **transmitter**. There are several different types of transmitters, including light emitting diodes (LEDs) and various types of lasers. Their purpose is to convert an electrical signal into an optical signal which can be carried by the fiber. This process is called **modulating** the source of light. For example, imagine a flashlight being switched on and off very quickly; the pattern of optical pulses forms a signal which carries information, similar to the way a telegraph or Morse code system operates. Turning a light source on and off in this way is called **direct modulation** or **digital modulation**. The light can also be adjusted to different levels of brightness or intensity, rather than simply being turned on and off, this is called **analog modulation**. Some types of light sources cannot be easily turned on or off; in this case, the light source may be left on all the time, and another device called an **external modulator** is used to switch the beam on and off (imagine a window shutter placed in front of a flashlight). The simplest type of modulation involves changing the intensity or brightness of the light; this is known as **amplitude modulation** (similar

1

to AM radio systems). It is also possible to modulate other properties of the light, such as its phase, frequency, or even polarization; however, these are not as commonly used in communication systems.

The **fiber** is a thin strand of glass or plastic which connects the light source to its destination (in the case of a communication system, it connects the transmitter to the receiver). Of course, there are other ways to carry an optical signal without using fibers which are beyond the scope of this book. For example, another type of **optical waveguide** can be made by layering polymers or other materials on a printed circuit board; these waveguides work on the same basic principles as an optical fiber. If the distances are fairly short, light can simply be directed through free space. Some laptop computers come equipped with infrared communication links that operate in this manner, much like a television remote control. This approach can also be used between two buildings if they are not too far apart (imagine two telescopes or lenses aimed at each other, with a light source on one side and a detector or receiver on the other side). This requires a clear line of sight between the transmitter and the receiver, as well as good alignment so that the beam does not miss the receiver; bad weather conditions will also affect how well this system works. We will focus our attention on the type of fibers most commonly used for communications, medical applications, and other related systems. The fiber often has layers of protective coatings applied to strengthen it and form an **optical cable**, in the same way that copper wires are often coated with plastic. Just as electrical connectors are required to plug a cable into a socket, **optical connectors** and sockets are also needed. Both fiber and copper cables may also be **spliced** to increase the distance, although the process is more difficult for fiber than for most copper systems. Most fibers used today are made from extremely high purity glass, and require special fabrication processes. We will describe the standard **fabrication equipment** needed to manufacture these fibers, such as **fiber drawing**, **chemical vapor deposition**, and **molecular beam epitaxy**.

In medical and lighting applications, the light is delivered to its destination to provide illumination or some other function. In a communication system, the light is delivered to an optical **receiver**, which performs the opposite function of the transmitter; it converts optical signals back to electrical signals in a communication system. A receiver is composed of an optical **detector** (sometimes called a **photodetector**) and its associated electronics. After traveling through the various components of a fiber optic link, the signal we are trying to detect will be corrupted by different types of noise sources. The receiver must be designed to separate a signal from this noise as much as possible. A simple approach would be to amplify the signal, and devices called **repeaters** and **optical amplifiers** may be inserted at various points along the link for this purpose. They can be used to extend the range of the system; however, bear in mind that amplification cannot solve all our problems, since it often increases the power of both the signal and the noise. There are also some types of noise which cannot be overcome simply by increasing the signal strength (for example, timing jitter on the digital data bits in a communication system). In some applications a transmitter and receiver are packaged together, this is called a **transceiver**.

Communication systems may also use other devices such as **multiplexers**, which combine several signals over the same optical fiber. Different wavelengths, or colors of light, can be carried at the same time over the same fiber without interfering with each other. This is called **wavelength division multiplexing (WDM)**; in later chapters we will discuss this in more detail. Optical fibers are often interconnected to form large **networks** which may reach around the world. It is also necessary to test and repair these networks using special tools. There are many tests of fiber optic systems that can be done with standard laboratory equipment. However, there are several pieces of equipment specific to fiber optics. An example of this is an **optical time domain reflectometer (OTDR)** that can non-destructively measure the location of signal loss in a link.

Non-communications applications of fiber optics include **medical technology**, **illumination**, **sensors**, and **controls**. Their sources differ dramatically from optical communications systems. Instead of requiring conversion from an electronic to an optical signal, the initial signal is often optical. Medical applications such as **fiberscopes** can use optical fibers to reach inside of the human body and create a picture. The signal is imaged directly onto a bundle of fibers. Additional fibers provide light to illuminate the image. Optical fibers can also be used in surgery; medical conditions can be diagnosed using fluorescence effects, and surgery can be performed using high intensity guided laser light. Optical fibers are also used in transducers and bio-sensors used for the measurement and monitoring of temperature, blood pressure, blood flow, and oxygen saturation levels [1].

There are other, non-medical, applications of imaging, illumination, sensors, and controls. Just as fiber lasers can be used for surgery, higher power systems are used in manufacturing for welding metal. Fiberscopes are also used for security and law enforcement, as well as for examining air ducts in buildings or other hard-to-reach places. Fiber optic lighting systems are used for emergency lighting, automotive lighting, traffic signals, signage, lighting sensors, and decorative lighting. Sometimes, their only detector is the human eye. There are many popular entertainment applications as well, including fiber optic decorations, artificial flowers, and hand-held bundles of fiber which show different colored lights using a white light source and a rotating color filter wheel.

As we can see, the field of fiber optics is very broad and touches many different disciplines. A comprehensive review is beyond the scope of this book; instead, we intend to describe the most common components used in fiber optic systems, including enough working knowledge to actually put these systems into practice (such as relevant equations and figures of merit).

1.1   Optical Fiber Principles

As noted earlier, we can direct light from one point to another simply by shining it through the air. We often visualize light in this way, as a bundle of **light rays** traveling in a straight line; this is one of the most basic approaches to light, called **geometric optics**. It is certainly possible to send useful information in this way (imagine signal fires, smoke signals, or ship-to-shore lights), but this requires an unobstructed straight line of light, which is often not practical. Also, light beams tend to spread out as they travel (imagine a flashlight spot, which grows larger as we move the light further away from a wall). We could make light turn corners using an arrangement of mirrors, but this is hardly comparable to the ease of running an electrical wire from one place to another. Also, mirrors are not perfect; whenever light reflects from a mirror, a small amount of light is lost. Too many reflections will make the optical signal too weak for us to detect. To fully take advantage of optical signaling, we need it to be at least as easy to use as a regular electrical wire, and have the ability to travel long distances without significant loss. These are the principal advantages of an optical fiber.

Instead of using mirrors, optical fibers guide light with limited loss by the process of **total internal reflection**. To understand this, we need to know that light travels more slowly through transparent solids and liquids than through a vacuum, and travels at different speeds through different materials (of course, in a vacuum the speed of light is about 300,000,000 m/s). The relative speed of light in a material compared with its speed in vacuum is called the **refractive index, n**, of the material.

For example, if a certain kind of glass has a refractive index of 1.4, this means that light will travel through this glass 1.4 times more slowly than through vacuum. The bending of light rays when they pass from one material into another is called **refraction**, and is caused by the change in refractive index between the two materials. Refractive index is a useful way to classify different types of optical materials; to give a few examples, water has a refractive index of about 1.33, most glass is around 1.5–1.7, and diamond is as high as 2.4. For now, we will ignore other factors that might affect the refractive index, such as changes in temperature. We can note, however, that refractive index will be different for different wavelengths of light (to take an extreme example, visible light cannot penetrate your skin, but X-rays certainly can). At optical wavelengths, the variation of refractive index with wavelength is given by the **Sellmeier equations**.

Total internal reflection is described by **Snell's Law**, given by Equation 1.1, which relates the refractive index and the angle of incident light rays. This equation is illustrated by Figure 1.1, which shows two slabs of glass with a ray of light entering from the lower slab to the upper slab. Here, n_1 is the index of refraction of the first medium, θ_1 is the angle of incidence at the interface, n_2 is the index of refraction of the second medium, and θ_2 is the angle in the second medium (also called the angle of refraction). Snell's Law states that

$$n_1 \sin \theta_1 = n_2 \sin \theta_2 \qquad (1.1)$$

Thus, we can see that a ray of light will be bent when it travels across the interface. Note that as we increase the angle θ_1, the ray bends until

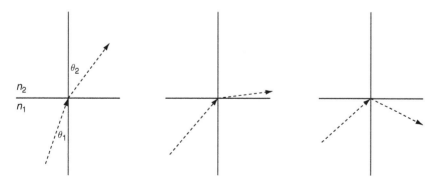

Figure 1.1 Illustration of Snell's Law showing how an incident light ray is bent as it travels from a slab of glass with a high refractive index into one with a lower refractive index, eventually leading to total internal reflection.

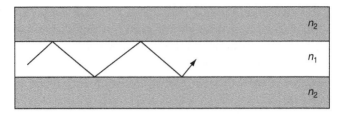

Figure 1.2 A sandwich of glass slabs with different indices of refraction used to guide light rays.

it is parallel with the interface; if we continue increasing θ_1, the light is directed back into the first medium! This effect is called **total internal reflection**, and it occurs whenever the refractive index of the first media is higher than the second media ($n_1 > n_2$). Now, imagine if we sandwich the high index glass between two slabs of lower index glass, as shown in Figure 1.2. Because total internal reflection occurs at both surfaces, the light bounces back and forth between them and is guided through the middle piece of glass. This is a basic optical waveguide. We can extend this approach by constructing a glass fiber, with a higher refractive index material surrounded by a lower refractive index material. Since the fiber **core** has a higher index of refraction than the surrounding **cladding**, there is a critical angle, θ_c, where incident light will be reflected back into the fiber and guided along the core. The critical angle is given by Equation 1.2,

$$\theta_c = \sin^{-1}\left(\frac{n_2}{n_1}\right) \tag{1.2}$$

Thus, if we make a fiber with a higher refractive index in the core and surround it with a lower index material as the cladding (as in Figure 1.3), then launch light waves at less than the critical angle, the

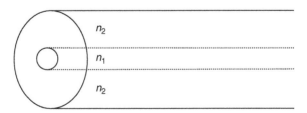

Figure 1.3 A cylindrical strand of glass with a higher index core forms an optical fiber.

light will be totally reflected every time it strikes the core-cladding interface. In this way, the light will travel down the fiber, following it around fairly tight corners or other bends. In practice, the fiber will also have an outer coating or jacket to add strength, since the fibers themselves are very small. For typical fibers used in communication systems, the refractive index difference between the core and the cladding is about 0.002–0.008 [2].

We can see that light from many different angles may enter the fiber, and take different paths as shown in Figure 1.4. Actually, this is a simplified picture of how light travels, which is useful to help us understand how an optical fiber works. For a more complete, accurate picture, we must realize that light is an electromagnetic wave, and an optical fiber is a dielectric optical waveguide. Light propagation follows **Maxwell's equations** which describe all types of electromagnetic fields. To describe an optical fiber, we can begin with these equations and derive the wave equation, which is a partial differential equation that can be solved using the boundary conditions in a cylindrical fiber. The mathematics is fairly complex, and yields expressions for the electromagnetic field of light in the fiber with a finite number of propagation modes. (See references [2] and [3] for a complete discussion of electromagnetic mode theory.) For our purposes, we will continue using the simplified model of fibers.

If the fiber shown previously has a large core diameter, light can enter from many different angles. Each of these will take a different path through the fiber. We can loosely call these different paths **modes** of propagation; if we use the more rigorous Maxwell's equations, they would correspond to the different propagation modes of electromagnetic waves in a bounded waveguide. Fibers with a large core can support several modes of propagation, and are called **multimode**. There are several standard fiber core diameters used for communication systems, including

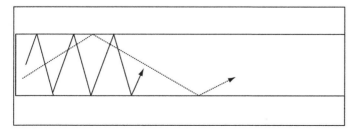

Figure 1.4 Light rays may take longer or shorter paths through an optical waveguide.

62.5 micron fiber (sometimes called "**FDDI grade**," since it was originally specified for an industry standard called Fiber Distributed Data
Interface), 50 micron fiber, and 100 micron fiber, which is less common and is not used for communication applications, but can be used
for imaging, sensors, and controls. The cladding diameter for communication fiber is typically 125 microns, and for other fibers it may be
around 140 microns. Fibers are often specified with both their core and
cladding sizes; for example, we can speak of "50/125" and "100/140"
fiber. Historically, 62.5/125 was designed in the 1980s for telecommunication systems to optimize both high bandwidth and long distance. It
was originally used with LED sources, but manufacturers have introduced new high-bandwidth 62.5/125 fiber that is compatible with some
types of lasers. The 50/125 fiber follows an earlier telecommunication
standard, but was reintroduced because it has higher bandwidth for laser
applications. Multimode fiber communication systems can be up to a few
kilometers long, without amplification [4]. Typically, multimode fibers
have an orange cover jacket. Recent vintage cables have begun to adopt a
convention of black connectors on 50 micron cables and beige connectors
on 62.5 micron cables [5]. However, standards on connector color are
particularly subject to change.

Multimode fiber distances are limited by interference between the
different modes, sometimes called **modal noise**. Also, we can see from
Figure 1.4 that different modes will travel different distances along the
fiber core, so even if they are launched into the fiber at the same time,
they will not emerge together. This is called **multipath time dispersion**.
We can calculate the amount of time dispersion per unit length of fiber
from

$$\text{Dispersion/length} = \frac{n_1 \Delta n}{n_2 c} \tag{1.3}$$

where n_1 and n_2 are the refractive indices of the core and cladding,
respectively, Δn is the difference between them, and c is the speed
of light. A strand of glass surrounded by air will guide light, since
the refractive index of glass is higher than the air; however, from this
equation, we can also see that it will have very high dispersion compared
with a fiber that uses another type of glass for cladding.

If we send a pulse of light into the fiber, this pulse will spread out as it
travels because of time dispersion. A high-speed digital communication
channel is just a sequence of pulses (for example, the digital "1" and "0"
of a computer signal can be represented as a light pulse turned on or off).
As we try to send higher and higher data rate signals, the pulses being

launched into the fiber become narrower and closer together. Eventually, dispersion will make these pulses spread out so that they overlap and it becomes impossible to tell one pulse from another. In this way, we see that increasingly higher communication data rates can only be supported over shorter and shorter lengths of multimode fiber. Put another way, the maximum useful **bandwidth** of the fiber, or the maximum **bit rate** for digital communications, is inversely related to the time dispersion. Multimode fiber is specified with a **bandwidth-distance product** written on the cable, for example 500 MHz-km; this means that an optical signal modulated at 500 MHz can travel up to 1 km before dispersion (and other effects) corrupt the signal beyond recognition. Similarly, an optical signal at half this speed (250 MHz) could go twice the distance (2 km). We have simplified this discussion, but clearly time dispersion between different fiber modes places a fundamental limit on the fiber bandwidth. On the other hand, the reason we are interested in optical fibers for communication is because their available bandwidth far exceeds that of copper wire.

As the fiber core grows smaller, fewer and fewer modes can propagate; this is shown in Figure 1.5. Fibers with a smaller core (8–10 microns) [6] which support essentially only one mode of propagation are called **single-mode**. They use laser sources and can go further unamplified distances (~ 20–50 km), because they do not have modal noise issues. Single-mode fibers do not have a bandwidth-distance product specification; their working distance is limited by other factors, such as signal loss or attenuation. Because of their smaller core size, alignment between two single-mode fibers or between a fiber and a transmitter/receiver

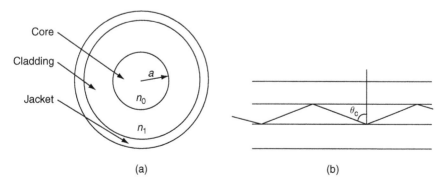

Figure 1.5 (a) Generic optical fiber design (b) path of a ray propagating at the geometric angle for total internal reflection.

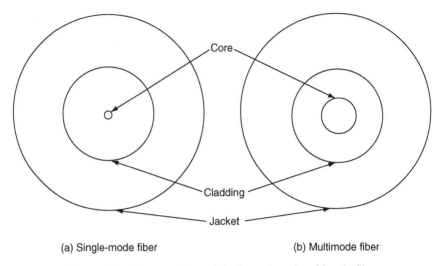

(a) Single-mode fiber   (b) Multimode fiber

Figure 1.6 A cross section of single-mode and multimode fiber.

is a difficult problem. While the actual price of single-mode fiber is not significantly different from multimode fibers, the price of related components is higher for this reason [7]. Typically, single-mode fibers have a yellow cover jacket and blue connectors, though as noted above connector colors are subject to change.

We see in Figure 1.6 a cross section of single-mode and multimode fiber with the core, cladding, and jacket labeled. Figure 1.7 is a chart of common optical fiber connectors used in communication systems. Connectors have different names based on their early history; for example, the common simplex connector is now widely known simply as an "SC" connector. Connectors are further differentiated by the mechanical method of joining the ends. For example, FC connectors use threaded fasteners, but SC connectors use a spring latch and bayonette styles. As can be seen, some connectors are common only for single-mode fiber, or only for multimode fiber, but some are also used for both. When considering the proper connector for a system, major concerns include **connection loss** (how much light is lost at the connector interface) and **repeatability** (consistent performance over a large number of mate–demate cycles). These can be expressed as either worst-case values or statistical distributions. For example, connector losses can be as high as 0.5 dB, which corresponds to an additional kilometer of fiber at a wavelength of 1.3 micrometers [2].

So far, we have only discussed optical fibers in which the core and the cladding have a constant index of refraction. If we want to minimize

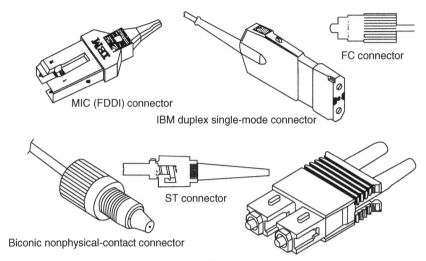

FC connector

MIC (FDDI) connector

IBM duplex single-mode connector

ST connector

Biconic nonphysical-contact connector

Fiber channel standard SC-duplex connector

Figure 1.7   Common fiber optic connectors: MIC (FDDI), ST, FC, ESCON, SC duplex, and biconic (nonphysical contact) [2].

the time dispersion effect, we might consider making the refractive index higher near the center of the fiber core, and lower near the edge of the core. This would make light rays near the center travel more slowly than rays near the edge, compensating for the time dispersion effect. The fiber core would then have a refractive index which changes along its diameter, similar to a lens; we can taper the index variation into the cladding, as well. When the refractive index of the core varies gradually as a function of radial distance from the fiber center, this is called a **graded-index fiber**. The previous case which we had described is called **step-index fiber**, since the core refractive index does not change and the index of refraction at the boundary between the core and the cladding is not continuous.

There are two basic types of single-mode **step-index fiber**—matched cladding and depressed cladding. **Matched cladding** is made from a single dielectric material. **Depressed cladding** has an inner and outer cladding material. The outer cladding material has an index of refraction that is higher than the inner cladding, but lower than the core. The advantage of this design is less loss due to chromatic and material dispersion, which will be discussed in a later chapter. Single-mode fiber also comes in graded index [8].

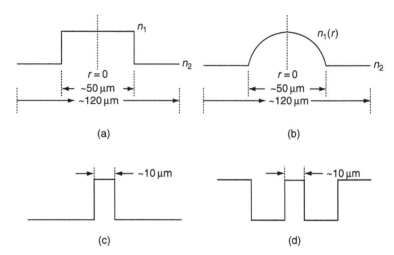

Figure 1.8 Refractive index profiles of (a) step-index multimode fiber (b) graded-index multimode fiber (c) match cladding single-mode fiber and (d) depressed cladding single-mode fiber.

Graded-index multimode fiber has the advantage that light rays travel at higher speeds in the outer layers, which have the greater path length. This results in equalizing the propagation time of the various modes, lowering modal dispersion. Both step-index and graded-index fibers are available in all typical multimode core sizes. This is the standard multimode cable used in most communication applications. Figure 1.8 shows the refractive profiles of (a) step-index multimode fiber, (b) graded-index multimode fiber, (c) match cladding single-mode fiber, and (d) depressed cladding single-mode fiber.

Most the fibers described above are made from doped silica glass, though some fibers are made from plastics. **Plastic optical fibers (POF)**, typically made from poly-methyl methacrylate (PMMA), have higher loss and are used for applications that do not require long transmission distances, such as medical instrumentation, automobile and aircraft control systems, and consumer electronics. One application of plastic fiber in a communications environment is optical loopbacks and wrap plugs used for transceiver testing. The short lengths required for a loopback means that attenuation is not an issue, sometimes it is even an advantage, because it reduces the risk of detector saturation. There is also some interest in using plastic fiber for the small office/home office (SOHO) environment.

Fibers are bundled into **cables** for transmission over large distances. There are two types of cable construction: (1) **tight-buffered** that is used for intrabuilding backbones that connect data centers, master controllers, and telecommunications closets and patch panels; (2) and **loose-tube** design that is used for outside plant environments and interbuilding applications.

1.2 Basic Terminology

When specifying a fiber link, **simplex** refers to a one-directional link; **duplex**, which uses two fibers, is required for two-way communication (it is possible to design duplex links using different wavelengths over a single fiber, which is a topic we will discuss in a later chapter).

The **numerical aperture (NA)** of a fiber measures the amount of light the fiber can capture. It refers to the maximum angle for which light incident on a fiber endface can still be refracted into the fiber and then undergo total internal reflection. It can be calculated by Equation 1.4.

$$n_1 \sin \theta \approx \theta = n_1{}^2 - n_2{}^2 = \text{NA} \tag{1.4}$$

where θ is the cone of the acceptance angle, NA is the numerical aperture, n_0 is the index of refraction of the medium where the light originates, n_1 is the index of refraction of core, and n_2 is the index of refraction of the cladding. Single-mode fibers have an NA of about 0.1, while the NA of multimode fibers vary from 0.2 to 0.3 [9].

The **coupling efficiency**, η, is defined as

$$\eta = \frac{P_{\text{acc}}}{P_{\text{input}}} = \frac{\pi \rho L_s A_f (\text{NA})^2}{P_{\text{input}}} \tag{1.5}$$

where P_{acc} is the power accepted by the fiber, and P_{input} is the input power, L_s is the radiance of the light source in watts per area and steradian, and A_f is the area of the fiber core, and ρ is a correction for the reduction due to reflection loss.

$$\rho = ((n_1 - n_0)/(n_1 + n_0))^2 \tag{1.6}$$

As can be seen, the area of the fiber core is very important in coupling efficiency. Figure 1.9 shows several examples of coupling efficiency between two optical fibers, including lateral misalignment, axial misalignment, and angular misalignment. Coupling efficiency can be improved by the use of a lens between the source and the fiber. The lens matches the output angle of the source to the acceptance angle of the fiber. In this

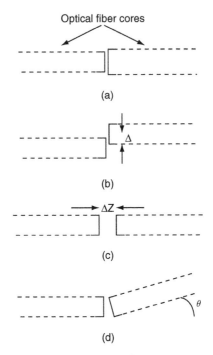

Figure 1.9   Different coupling cases (courtesy of Academic Press).

case, the previous formula must be multiplied by the lens magnification factor **M**.

From a transmission point of view, the two most important fiber parameters are **bandwidth** and **attenuation**. The fundamental reason we are using fiber instead of copper cable is the increased **bandwidth**. Bandwidth is the difference between the highest and the lowest frequency information that can be transmitted by a system. A higher bandwidth implies a greater capacity for a channel to carry information. Bandwidth is tested using a very fast laser and a sensitive receiver. Software analyzes the difference between the input and the output pulses, and calculates the bandwidth of the fiber.

Bandwidth is also design dependent—for example, the bandwidth of a step-index multimode fiber ($\sim$125 MHz) is lower than for a graded-index multimode fiber ($\sim$500 MHz). Table 1.1 shows bandwidth variations on two 100 m lengths of FDDI-grade fiber with different mid-span connections—connector, mechanical splice, and fusion splice. As noted earlier, multimode fibers are generally specified by their bandwidth in a 1 km length, which ranges from 100 MHz to 1 GHz. For example, a

Table 1.1 **Bandwidth variations with various termination types**

	Connector	Mechanical splice	Fusion splice
Bandwidth Range	1.28–1.34 GHz	1.09–1.66 GHz	0.83–1.41 GHz

62.5/125 fiber cable jacket may be labeled as "500 MHz-km," meaning that over a distance of 1 km the fiber can support a minimum bandwidth of 500 MHz. If the same fiber is required to carry a signal at twice this rate, or 1 GHz, we would only be able to use a link half as long, or 0.5 km, before the signal become unrecognizable due to bandwidth limits in the channel.

Bandwidth is inversely proportional to **dispersion** (pulse spreading), with the proportionality constant dependent on pulse shape and how bandwidth is defined. We have already mentioned that single-mode fiber is used at longer distances because it does not have **intermodal dispersion**. Other forms of dispersion are **material dispersion** (because the index of refraction is a function of wavelength) and **waveguide dispersion** (because the propagation constant of a fiber waveguide is a function of wavelength). Detailed mathematical descriptions of these properties are beyond our scope.

Attenuation is a decrease in signal strength caused by **absorption**, **scattering**, and **radiative loss**. The power or amplitude loss is often measured in decibels (dB), which is a log scale.

$$dB = 10 \ \log \left(\frac{\text{power level in W}}{1 \text{ W}} \right) \tag{1.7}$$

If the power levels are defined per mW, the same equation is used but the scale is now decibels/milliwatt or dBm. Attenuation in an optical fiber is a function of the operating wavelength. Typically, silica glass fibers have an attenuation minimum near 1.5 micron wavelength (about 0.25 dB/km), which is commonly used for long haul telecommunications and WDM applications. The attenuation is somewhat higher at 1.3 micron wavelength (around 0.5 dB/km), although this range is often used for data communication and other applications with maximum distances less than about 10 km because silica fiber has a dispersion minima at this wavelength. Thus, applications whose performance is likely to be dispersion limited, rather than loss limited, should operate at this point. The loss is significantly higher (around 3–4 dB/km) near 850 nm wavelength, however low cost, high reliability optical sources are readily available at this wavelength, making it popular for distances less than 1 km. There are also a number of specialty optical fibers available which have been optimized to have lower loss and higher bandwidth at different wavelengths.

Radiative loss is minimized by designing the fibers with sufficiently thick cladding (~ 125 microns) and minimizing bends (the so-called **bending loss** is a form of radiative loss). Fiber optic materials are made of very high purity materials (less than a few parts per million) to limit **scattering loss**. **Absorption**, however, is a material constraint, a result of the so-called "Rayleigh scattering" from the glass structure. This leads to an inherent loss for the "best" fibers of

$$\alpha = \frac{B}{\lambda^4}, \text{ with } B = 0.9 \text{ (dB/km) } \mu m^4 \tag{1.8}$$

where α is absorption and λ is wavelength in microns.

A common measurement of attenuation is the cutback method. Fiber transmission as a function of wavelength $P_A(\lambda)$ is measured for the full length of the fiber. The fiber is then cut ~ 1–2 mm away from the input end, and the transmission is measured once again. If this second measurement is called $P_B(\lambda)$, then the fiber attenuation constant α_f may be calculated from

$$\alpha_f = 10 \log \left[\frac{P_A(\lambda)}{P_B(\lambda)} \right] \tag{1.9}$$

Note that this is a destructive method of testing. The cutback method is typically used during fiber manufacturing. To measure attenuation in the field or laboratory, a comparison is done with a reference cable.

1.3   Single-Mode Fiber

Single-mode fiber is the most common choice for long distance communication. It does not have modal dispersion, which distorts the signal pulse at long distances. It still experiences chromatic and material dispersion, and attenuation.

It can be shown [9] that the condition for single-mode propagation in a fiber is that the **normalized frequency** V be less than 2.405, where

$$V = 2\pi a \frac{NA}{\lambda} \tag{1.10}$$

and a is the core radius (from Equation 1.5, $A_f = \pi a^2$), λ is the free space wavelength, and NA is the numerical aperture.

For good single-mode performance, the normalized frequency V should lie between about 1.8 and 2.4. For values above 2.405, the fiber is multimode. Normalized frequency and several other concepts in this section may be derived from Maxwell's equations, though this derivation is beyond the scope of this book.

The **normalized propagation constant**, b, which is the phase constant of the mode traveling through the fiber, is defined as

$$b = \frac{((\frac{\beta^2}{k^2}) - n_2^2)}{(n_1^2 - n_2^2)} \tag{1.11}$$

where β is the phase constant of the particular mode, k is the propagation constant in vacuum, n_1 is the index of refraction of core, and n_2 is the index of refraction of the cladding.

If b is zero, then the mode cannot propagate through the fiber. The wavelength for which this condition holds is known as the **cutoff wavelength**. It is the result of the solution of the electromagnetic wave equations propagating down a waveguide.

The cutoff wavelength λ_{co} can be calculated by

$$\lambda_{co} = \frac{2\pi a \text{NA}}{V_{co}} \tag{1.12}$$

The normalized propagation constant b as a function of V is shown in Figure 1.10. The cutoff frequency can be found on the graph when

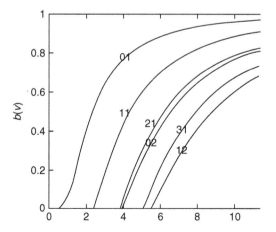

Figure 1.10   Cutoff frequencies for the lowest order LP modes (courtesy of Academic Press).

Table 1.2 Cutoff frequencies of various $LP_{\ell m}$ modes in a step-index fiber (courtesy of Cambridge University Press)

$\ell = 0\,modes$	$J_1(V_c) = 0$	$\ell = 1\,modes$	$J_o(V_c) = 0$
Mode	V_c	Mode	V_c
LP_{01}	0	LP_{11}	2.4048
LP_{02}	3.8317	LP_{12}	5.5201
LP_{03}	7.0156	LP_{13}	8.6537
LP_{04}	10.1735	LP_{14}	11.7915
$\ell = 2\,modes$	$J_1(V_c) = 0;\ V_c \neq 0$	$\ell = 3\,modes$	$J_2(V_c) = 0_i;\ V_c \neq 0$
LP_{21}	3.8317	LP_{31}	5.1356
LP_{22}	7.0156	LP_{32}	8.4172
LP_{23}	10.1735	LP_{33}	11.6198
LP_{24}	13.3237	LP_{34}	14.7960

$b = 0$. Table 1.2 shows a list of cutoff frequencies of various modes in step-index fiber.

In Table 1.3, we see typical specifications of Single-Mode Glass Optical Fiber (per Consultative Committee of International Telephone and Telegraph (CCITT) Recommendation G.652). A fiber of this type should be able to transmit data unrepeated over many kilometers.

Table 1.3 Typical specifications of single-mode glass optical fiber (per Consultative Committee for International Telephone and Telegraph (CCITT) Recommendation G.652)

Parameter	Specification
Core diameter (mode field diameter)	9–10 μm
Cladding diameter	125 μm
Cutoff wavelength	1100–1280 nm
Bend loss at 1550 nm	<1 dB for 100 turns around a 7.5 cm diameter mandrel
Dispersion	<3.5 ps/nm km between 1285 and 1330 nm <6 ps/nm km between 1270 and 1340 nm <20 ps/nm km at 1550 nm
Dispersion slope	<0.095 ps/nm² km

1.4   Multimode Fiber

While it is possible to use step-index multimode fiber, typical communications applications use graded-index fiber to reduce intermodal distortion. The graded-index profile can be generated with the expressions

$$n^2(r) = n_1{}^2[1 - (\text{NA})^2(r/a)^q] \qquad \text{for } 0 < r < a,$$
$$n^2(r) = n_2{}^2 \qquad\qquad\qquad \text{for } r > a \qquad (1.13)$$

when q is called the profile exponent and r is the radial distance away from the center of the fiber core, and n_1 is the index of refraction in the center of the fiber and n_2 is the index of refraction of the cladding. The optimum profile occurs for q slightly less than 2, which results in the minimal dispersion.

The total number of modes that can propagate in a multimode fiber N is given by the following expression, valid for large V numbers only:

$$N = 1/2 \, q \, V^2/(q + 2) \qquad (1.14)$$

where q is the profile exponent and V is the normalized frequency. For the special case of step-index fiber, q tends to infinity and $N = V^2/2$. In Table 1.4, we show typical specifications Δ of multimode glass-optical fiber.

Table 1.4   **Typical specifications of multimode glass-optical fiber**

Type of fiber	Graded index with glass core and cladding
62.5/125 micron multimode fiber	
Core diameter	$62.5 \pm 3.0\,\mu\text{m}$
Core noncircularity	6% maximum
Cladding diameter	$125 \pm 3.0\,\mu\text{m}$
Cladding noncircularity	2% maximum
Core and cladding offset	$3\,\mu\text{m}$ maximum
Numerical aperture	0.275 ± 0.015
Minimum modal bandwidth	500 MHz ∗ km at ≤2 km at 1300 nm or
	160 MHz ∗ km at 850 nm
Attenuation	1.0 dB/km at 1300 nm or
	4.0 dB/km at 850 nm

Table 1.4   (continued)

Type of fiber	Graded index with glass core and cladding
50/125 micron multimode fiber	
Core diameter	$50.0 \pm 3.0\,\mu$m
Core noncircularity	6% maximum
Cladding diameter	$125 \pm 3.0\,\mu$m
Cladding noncircularity	2% maximum
Core and cladding offset	$3\,\mu$m maximum
Numerical aperture	0.200 ± 0.015
Minimum modal bandwidth	800 MHz * km at $\leq$2 km at 1300 nm or
	500 MHz * km at $\leq$1 km at 850 nm
Attenuation	0.9 dB/km at 1300 nm or
	3.0 dB/km at 850 nm

1.5   Fiber Bragg Gratings (FBGs)

Fiber Bragg gratings are spectral filters fabricated within segments of optical fiber. Like any other diffraction grating, they rely on the Bragg Effect (constructive interference from a diffraction grating) to reflect light over a narrow wavelength range and transmit all other wavelengths, but they also can be designed to have more complex spectral responses. Many uses exist for FBGs in today's fiber communications systems, so although they are complex to design we will briefly describe their properties.

When an optical fiber is exposed to ultraviolet light, the fiber's refractive index is changed; if the fiber is then heated or annealed for a few hours, the index changes can become permanent. The phenomenon is called **photosensitivity** [10, 11]. In germanium-doped single-mode fibers, index differences between 10^{-3} and 10^{-5} have been obtained. Using this effect, periodic diffraction gratings can be written in the core of an optical fiber. This was first achieved by interference between light propagating along the fiber and its own reflection from the fiber endface [12]; this is known as the internal writing technique and the resulting gratings are known as **Hill gratings**. Another approach is the transverse holographic technique in which the fiber is irradiated from the side by two beams which intersect at an angle within the fiber core. Gratings can also be written in the fiber core by irradiating the fiber through a phase mask with a periodic structure. These techniques can be used to write FBGs in the fiber core; such gratings reflect light in a narrow bandwidth centered around the Bragg wavelength, λ_B, which is given by

$$\lambda_B = 2N_{eff}\Delta \qquad (1.15)$$

where Δ is the spatial period, or pitch, of the periodic index variations and N_{eff} is the effective refractive index for light propagating in the fiber core. There are many applications for FBGs in optical communications and optical sensors, such as tapped optical delay lines, filters, multiplexers, optical strain gauges, and others (an extensive review is provided in references [13, 14]). FBGs function in reflection, while many applications require transmission effects; this conversion is accomplished using designs such as an optical circulator, a Michelson or Mach-Zender interferometer, or a Sagnac loop [14]. FBGs can be used to multiplex and demultiplex wavelengths in a WDM system, or to fabricate add/drop filters within the optical fibers that offer very low insertion loss; they can also be used in various dispersion compensation schemes [15].

Fiber Bragg gratings can also be used for remote monitoring applications, such as measuring the structure and function of an oil well to prevent hazards or malfunctions. The FBG is one of the most common fiber sensors. A change in external conditions – temperature or mechanical force, for example – will expand or compress the FBG, altering the refractive index period, and changing the wavelength reflected. By using either a broad wavelength source coupled with a Michelson interferometer, or a tunable laser coupled with an absolute wavelength reference, the change in temperature or pressure can be measured [16, 17, 18].

1.6   Plastic Optical Fiber

Plastic optical fibers have larger cores (120–1000 microns) than standard multimode fiber, which makes it easier to align connectors. They are easy to install, but have high transmission losses (often several dB/km) which limit their applications to fairly short distances. Often their source is visible light, while infrared light is more common for communications. Graded-index POFs can be used for short distance links ($<100\,$m) although maximum link lengths are considerably shorter than glass multimode fibers. An application of POF is home entertainment system networking – it is well suited to red light sources (650 nm) which can carry DVD signals [19]. When POF is designed for consumer use in the home, both square connectors and round (Optical Mini Jack) connectors are used in portable media players. The defining standard is S/PDIF (Sony/Phillips Digital InterFace) or IEC 60958 [20].

In Table 1.5 we see typical specifications of POF used for communication applications. Various types of plastic fibers and connectors have been proposed as standards for communications applications. For example, Lucina graded-index POF is made of transparent fluorpolymer, CYTOP. It is available in both single-mode and multimode versions, and can potentially support over 1 Gbit/s up to 500 m; attenuation is about 50 dB/km at 850 nm, and it has a bandwidth of 200–300 MHz-km [21].

A list of general lighting applications of POF includes decorative lighting for displays, toys and models, museum display lighting, jewelry case/showcase display lighting, walk-away guidance lighting, commercial display lighting, photoelectric controls, elevators and conveyors, automotive dashboard instrumentation lighting, seat pocket, glove box and trunk lighting, and swimming pool and sauna lighting. In these applications, POF is lightweight, low cost, and has high attenuation and easy termination is essential [22].

Fiberoptic endoscopes, also known as fiberscopes, are used for medical imaging and security applications – POF is more flexible than glass fiber, and their large cores are well suited to imaging. But for different applications, both borosilicate glass and fiber are used. The flexible scope consists of a polyamide outer sheath, illumination fiber, and an image bundle with an objective lens system at the distal end. Fiberscopes are discussed in more detail in a later chapter.

When using optical fibers for industrial lighting applications, a typical source of light is a quartz halogen projection lamp. Coupling this with a standard plastic or borosilicate fiber optic cable can illuminate a variety

Table 1.5 **Typical specifications of plastic optical fiber**

Parameter	*Specification*
Core diameter	980 μm
Cladding diameter	1000 μm (1 mm)
Jacket diameter	2.2 mm
Attenuation (at 850 nm)	<18 dB/100 m (180 dB/km)
Numerical aperture	0.30
Bandwidth (at 100 m)	Step index: 125 MHz
	Graded index: 500 MHz

of surgical instruments, headlights, and fiberscopes. The end ports are determined by the receptacle port of the lightsource, and the connector port of the headlight, scope or instrument. In Figure 1.11, we show a variety of typical connectors.

1.7 Cables

Optical fibers are bundled in cables for distribution and installation. There are two typical cable designs: (1) loose-tube cable, used in the majority of outside-plant installations in North America; and (2) tight-buffered cable, primarily used to distribute fibers inside buildings [23] (Figure 1.12).

Loose-tube cables are used primarily in aerial, duct, and direct buried applications. Each color-coded plastic buffer tube (see TIA/EIA-598-A "Optical Fiber Cable Coloring Coding") houses and protects up to12 fibers, and can be bundled together to hold up to 200 fibers. These fibers are surrounded by a gel to impede water penetration. These tubes are wound around a cable core which typically uses a central member of dielectric aramid yarn, fiberglass, or steel to provide tensile strength. The cable has an outer plastic jacket extruded over the core. If armoring is required, a corrugated steel tape is formed around a single jacketed cable with an additional jacket extruded over the armor [24] (see Figure 1.13). When using bundles of fiber cables, the optical specifications for cables should include the maximum performance of all fibers, not just typical fibers, over the operating temperature range and life of the cable.

Tight-buffered cables put the buffering material directly in contact with the fibers. This design provides a rugged cable structure to protect individual fibers during handling, routing, and connectorization. Yarn strength members keep the tensile load away from the fiber. The cables are thinner and the fibers are more accessible than in the loose-tube design. In fact, a zip-cord may be applied between strength yarns and the outer jacket to facilitate jacket removal [25] (Figure 1.14).

Compared to copper cables, fiber optic connectors can be relatively difficult to install. If it is possible, field installation should be avoided by buying pre-connectorized cables that are made and tested by the manufacturer. The other approach is to contract a professional fiber optic installer. This service is currently more expensive than a typical electrical contractor. The fiber optic installer who is placing connectors on fiber has to work with hair-thin glass fibers (which can form sharp glass shards than find their way under the skin), mix and apply epoxy, polish the end

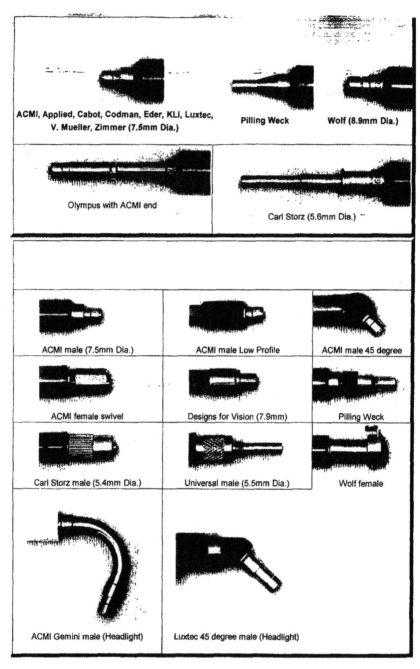

Figure 1.11   Connectors used for fiber optic lighting applications.

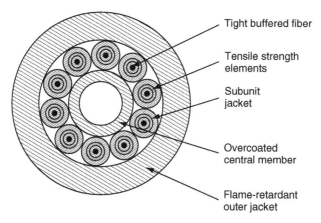

Figure 1.12   Tight-buffered fan-out cable.

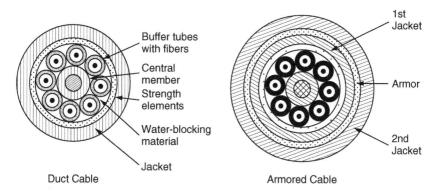

Figure 1.13   Stranded loose-tube cable.

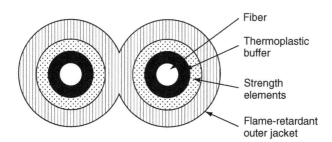

Figure 1.14   Two-fiber zipcord cable.

of the fiber to a mirror finish, and align it with micron precision. This is often done when confined to a tight space.

Another approach is to splice the fiber to a short manufactured cable with a connector. This can be easier, because the ends do not require a mirror finish. This pigtail can be purchased as "no-cure, no-polish connectors," or the installer can cleave a cable. While this approach can be up to five times more expensive than attaching a plain connector, labor is costly and on-site polishing can be tricky. While splices often connect different types of cable—for example, outdoor armored loose tube fiber with indoor tight-buffered cable—it is important to splice the fiber to the same size of fiber. For example, 62.5/125 fiber should be spliced to another 62.5/125 fiber and not to a 50/125 fiber.

There are two major types of field splices. **Fusion splicing** aligns the core of two clean (stripped of coating) cleaved fibers and fuses the ends together with an electric arc. Alignment is done using a fixed V-groove to hold the fibers, using a microscope to visually align the fiber profiles or by injecting light through the splice and adjusting the fiber position until the light intensity is maximized. **Mechanical splicing** holds two fibers end to end in a tube with index matching gel. This splice relies on the alignment of the outer diameter of the fibers, making the accuracy of the core/cladding concentricity critical to achieving low splice losses.

References

[1] Lee, B. (2004) Fiber Optics Tutorial, Nanoptics (http://www.nanoptics.com/tutorial.htm).

[2] Ulf L. Osterberg (2002) "Optical Fiber, Cable and Connectors", in *Handbook of Fiber Optic Data Communication*, C. DeCusatis (ed.), Chapter 1, pp. 3–38, Academic Press.

[3] Brown, T.G. (2001) "Optical Fibers and Fiber-Optic Communications", in *Handbook of Optics*, M. Bass, *et al.* (ed.), pp. 1.6–1.9. McGraw-Hill.

[4] Hayes, J. and LeCheminant, G. (2002) "Testing Fiber Optic Local Area Networks (LANs)", in *Handbook of Fiber Optic Data Communication*, C. DeCusatis (ed.), Chapter 9, p. 336, Academic Press.

[5] DeCusatis, C. (2002) "Fiber Optic Link Requirements", in *Handbook of Fiber Optic Data Communication*, C. DeCusatis (ed.), Chapter 7, p. 268, Academic Press.

[6] Fiber Optics Tutorial, International Engineering Consortium 2004 (http://www.iec.org/online/tutorials/fiber_optic/topic02.html).

[7] Clement, D.P., Lasky, R.C., Baldwin, D. (2002) "Alignment Metrology and Techniques", in *Handbook of Fiber Optic Data Communication*, C. DeCusatis (ed.), Chapter 19, p. 703, Academic Press.

[8] Introduction to Fiber Optics, Naval Electrical Engineering Training Series, volume 24, Integrated Publishing (http://www.tpub.com/neets/book24/index.htm).

[9] Jacobs, I. (2001) "Optical Fiber communication Technology and System Overview", in *Handbook of Optics*, M. Bass (ed.), *et al.*, p. 2.2, McGraw-Hill.

[10] Hill, K. (2000) "Fiber Bragg Gratings", *Handbook of Optics*, Chapter 9, vol. IV, OSA Press.

[11] Poulmellec, B., Niay, P., Douay, M., *et al.* (1996) "The UV-induced refractive index grating in Ge:SiO_2 Preforms: Additional CW experiments and the macroscopic origins of the change in index", *J Phys. D.: App. Phys.*, 29:1842–1856.

[12] Hill, K.O., Malo, B., Bilodeau, F., *et al.* (1993) "Photosensitivity in optical fibers", *Ann. Rev. Mater. Sci.*, 23:125–157.

[13] Kawasaki, B.S., Hill, K.O., Johnson, D.C., *et al.* (1978) "Narrowband Bragg reflectors in optical fibers", *Opt. Lett.*, 3:66–68.

[14] Hill, K. and Meltz, G. (1997) "Fiber Bragg grating technology: fundamentals and overview", *J. Lightwave Tech.*, 15L, 1263–1276.

[15] DeCusatis, C. (2002) "Optical Wavelength Division Multiplexing", in *Fiber Optic Data Communications: Technological Trends and Advances*, C. DeCusatis (ed.), Chapter 5, pp. 169–170, Academic Press.

[16] Anne L. Fischer (2003) "Wavelength Monitoring of Sensors-Two Approaches", *Photonics Spectra*, October.

[17] Tony, A. and William, G. (2003) "Michelson Interferometry Interrogates Fiber Bragg Grating Sensors", *Photonics Spectra*, October.

[18] Christopher J. Myatt and Nicholas G. Traggis (2003) "Tunable Laser Wavelength Calibration for Fiber Sensing", *Photonics Spectra*, October.

[19] Lucent Technologies announces new plastic optical fiber for specialty fiber applications, Press Release, 9 June 2000, Lucent Technologies (http://www.lucent.com/press/0600/000609.bla.html).

[20] Plastic Optical Fiber Data Sheet, Sharpe Microelectronics, 2005 (http://www.sharpsma.com/sma/Products/Prod-frame/opto.htm?main=http://www.sharpsma.com/sma/products/opto/FiberOpticDevices.htm).

[21] DeCusatis, C. and John, F. (2002) "Specialty Fiber Optic Cables", *Fiber Optic Data Communications: Technological Trends and Advances*, C. DeCusatis (ed.), Chapter 4, p. 123, Academic Press.

[22] PMMA Plastic Optical Fiber Product Reference Sheet, Techman International Corporation 1997–2004 (http://www.techmaninc.com/PDF/PMMA%20Fiber%20Sheet.pdf).

[23] The Basics of Fiber Optic Cable, 2005 Data Connect Enterprise (http://www.data-connect.com/Fiber_Tutorial.htm).

[24] Specification for Standard Loose Tube Cables, Kamloops Community Network, City of Kamloops, BC, CA, 2003 (http://www.city.kamloops.bc.ca/kcn/pdfs/Appendix%20I.pdf).

[25] Corning Cable Systems Generic Specification for Tight Buffer Optical Fiber Cables for Indoor Distribution Applications, 2003 (http://www.corningcablesystems.com/web/library/AENOTES.NSF/$ALL/PGS007/$FILE/PGS007.pdf).

Chapter 2 | Transmitters

In this chapter we will discuss optical transmitters for fiber optic systems. This includes basic fiber optic communications light sources, such as LEDs, different types and wavelengths of lasers, and optical modulators. Our focus in this chapter is primarily on communication and sensor systems. Certain types of optical transmitters have become optimized for fiber optic communications, and other sources used for illumination and imaging have very different specifications and operating principles. While we discussed the many types of optical fiber in a single chapter, because a number of different applications use the same type of fiber, we cannot do the same with light sources; a fiber optic link is fundamentally different from a physician's fiberscope. This chapter describes the type of transmitters used in fiber optic links; sources for medical, industrial and lighting applications have different characteristics and will be treated in a separate chapter.

Following a review of basic terminology, we will discuss LEDs, lasers, and optical modulators. LEDs are frequently used for multimode fiber optic links. The two types of LEDs, surface emitting LEDs and edge emitting LEDs (ELEDs), have high **divergence** of the output light (meaning that the light beams emerging from these devices can spread out over 120° or more); slow rise times (>1 ns) that limit communication to speeds of less than a few hundred MHz; and output powers on the order of 0.1–3 mW. However, LEDs are less temperature sensitive than lasers and are very reliable.

For higher data rates, laser sources at short wavelengths (780–850 nm) are typically used with multimode fiber, and longer wavelengths

(1.3–1.5 microns) with single-mode fiber. Typical lasers used in communications systems are **edge emitting** lasers (light emerges from the side of the device), including double heterostructure (DH), quantum well (QW), strained layer (SL), distributed feedback (DFB), and distributed Bragg reflector (DBR). Vertical cavity **surface emitting** lasers (VCSELs) have been introduced for short wavelength applications, and are under development for longer wavelengths; fibers may couple more easily to these surface emitting sources. Compared with LEDs, laser sources have much higher powers (3–100 mW or more), narrower spectral widths (<10 nm), smaller beam divergences (5–10 degrees), and faster modulation rates (hundreds of MHz to several GHz or more). They are also more sensitive to temperature fluctuations and other effects such as back reflections of light into the laser cavity.

While direct modulation is common for LEDs, it causes various problems with lasers, including turn-on delay, relaxation oscillation, mode hopping, and frequency chirping. External modulators used with continuous wave (CW) diode lasers are one possible solution (e.g. **lithium niobate modulators** based on the electro-optic effect, used in a Y-branch interferometric configuration). **Electroabsorption modulators**, made from III-V semiconductors, are also used because they allow for more compact designs and monolithic integration. **Electro-optic** and **electrorefractive** modulators are another option, which use phase changes, rather than absorption changes, to modulate light.

2.1   Basic Transmitter Specification Terminology

There are several common features which we will find in a typical optical transmitter specification, regardless of the type of communication system involved. Every transmitter specification should include a physical description with dimensions and construction details (e.g. plastic or metal housing) and optical connector style. Specifications should also include a list of environmental characteristics (temperature ranges for operation and storage, humidity), mechanical specifications, and power requirements. Device characteristic curves may also be included, such as optical power output vs. forward current, or spectral output vs. wavelength.

A table of **absolute maximum ratings** will list properties such as the highest temperature for the case and device, as well as the maximum temperature to be used during soldering, the maximum current in forward or reverse bias, and the maximum operation optical power and current.

Care should be taken with more recent vintage devices, which may use environmentally friendly lead-free solders and thus require higher soldering temperatures. Temperature is a very significant factor in determining the operation and lifetime (reliability) of a semiconductor optical source. This is one reason why many optical transceivers for communication systems are being developed as modular, pluggable components which do not require soldering to a printed circuit card.

The **spectral width**, $\Delta\lambda$, is a measure of the range of optical frequencies or the spectral distribution emitted by the source, given in nanometers. This may be calculated from the spectrum as the full width at half maximum (FWHM), which happens to exactly define the **spectral bandwidth**. Other measurements may include the root mean square (RMS) spectral width, which is a statistical value. LEDs will have a very broad spectrum compared with lasers; short wavelength lasers will have a broader spectrum than long wavelength lasers (this is another way of saying that the short wavelength lasers are less **coherent**).

The **center wavelength** is the midpoint of the spectral distribution (the wavelength that divides the distribution into two parts of equal energy). It is given in nanometers or micrometers (microns). The **peak wavelength**, λp, is the spectral line with the highest output optical power. The center wavelength and the peak wavelength are usually both listed on most specifications, although occasionally red LEDs which were designed to be coupled with plastic fiber leave out the central wavelength. The terminology **spectral bandwidth** and **spectral width** are used interchangeably in these specifications.

The **temperature coefficient of wavelength** measures how the temperature affects the output power at a specific wavelength. It is expressed in units of nanometers per degree Celsius. The **temperature coefficient of the optical power** is a broadband measurement, which is expressed in a percentage per degree Celsius.

The **coupling efficiency** between a source and a fiber is the ratio of the input power (the available power from the source), to the power transferred to the fiber or other component, converted to percent. A related value is the **fiber coupled power**, **Poc**, which is the coupling efficiency multiplied by the radiant power; this is given in units of either mW or dBm. Different values for the fiber coupled power are normally listed for different sized cores, especially the standard multimode fiber core diameters of 50 micron, 62.5 micron and 100 micron or a standard single-mode fiber of 9 micron.

The **rise/fall time** is a measurement of how fast the source can be directly modulated, or how fast the light intensity can be ramped up and down, usually in nanoseconds. A typical measurement is the 10/90 rise or fall time, which is the time during which the source output rises from 10 to 90 percent of its peak value, or falls from 90 to 10 percent. Sometimes the 80/20 rise and fall times are specified instead. The **data rate** for serial transmission is simply the number of bits per second which can be transmitted. This should not be confused with the modulation frequency, measured in cycles per second, which is half the bit rate. For example, consider a square wave which is being transmitted by an optical source; if the frequency is 100 MHz, then a complete square wave is transmitted every 0.01 microseconds. However, this is enough time to transmit 2 data bits, so the bit rate would be 200 Mbit/second. The time required to transmit one bit is called the **unit interval**, as expressed in bits/second; it is often specified in communication standards.

2.1.1 LASER SAFETY

Generally speaking, fiber optic systems are no more dangerous to work with than any other type of electronics system if the user is aware of the potential hazards and takes nominal precautions to avoid them. Whenever designing or working with optical sources, eye safety is a significant concern. All types of optical sources, including lasers, LEDs, and even white light sources can cause vision problems if used incorrectly (e.g. if they are viewed directly through a magnifying lens or eye loop). In particular, fiber optic systems typically use infrared light which is invisible to the unaided eye; under the wrong conditions, an infrared source of less than 1 mW power can cause serious eye injury. Optical safety is not limited to viewing the source directly, but may also include viewing light emerging from a open connector, especially if a lens is used as noted earlier. In the United States, the Department of Health and Human Services (DHHS) of the Occupational Safety and Health Administration (OSHA) and the U.S. Food and Drug Administration (FDA) defines a set of eye safety standards and classifications for different types of lasers (FDA/CDRH 21 CFR subchapter J). International standards (outside North America) are defined by the International Electrotechnical Commission (IEC/CEI 825) and may differ from the standards used in North America. The basic standard for the safe use of lasers was first issued in 1998 by the American National Standards Institute; since then, several updates and additional standards have been released (ANSI standard Z136.X, where X refers to one of six existing standards for lasers in different environments). Copies of these standards are available from various sources, including the Laser Institute of America (LIA), which also offers laser safety training. There may be additional state or local regulations governing safe use of lasers as well (e.g. New York State Code Rule 50 requires all laser manufacturers to track their primary components using a state-issued serial number). Anyone working with fiber optics on a regular basis should be familiar with the different types of laser safety classifications and how they apply to a specific product. Laser safety training is required by law in order to work with anything other than class 1 products (see Table 2.1 for a list of laser safety classifications). All lasers other than class 1 (inherently safe) require a safety label; their classification depends on factors including the optical wavelength, power level, pulse duration, and safety interlocks built into their packaging. Fiber optic practitioners should be familiar with the safety requirements for systems they will be using.

In addition, practitioners should be aware of all standard safety precautions for handling fiber optic components. For example, cleaving and splicing of glass fibers can produce very small, thin shards of glass

Table 2.1   **ANSI Laser Safety Classifications**

Class 1

- inherently safe
- no viewing hazard during normal use or maintenance
- no controls or label requirements (typically 0.4 microwatts or less output power)

Class 1M

- inherently safe if not viewed through collecting optics
- designed to allow for arrays of optical sources which may be viewed at the same time

Class 2

- normal human eye blink response or aversion response is sufficient to protect the user
- low power visible lasers (<1 mW continuous operation)
- in pulsed operation, warning labels are required if power levels exceed the class 1 acceptable exposure limits for the exposure duration, but do not exceed class 1 limits for a 0.25 second exposure

Class 2A

- warning labels are required
- low power visible lasers that do not exceed class 1 acceptable exposure limits for 1000 seconds or less and systems which are not designed for intentional viewing of the beam

---

Table 2.1 (*continued*)

Class 3A

- normal human eye blink response or aversion response is sufficient to protect the user, unless laser is viewed through collecting optics
- requires warning labels, enclosure/interlocks on laser system, and warning signs at room entrance where the laser is housed
- typically 1–5 mW power

Class 3B

- direct viewing is a hazard, and specular reflections may pose a hazard
- same warning label requirements as class 3A plus power actuated warning light when laser is in operation
- typically 5–500 mW continuous output power, <10 joules per square centimeter pulsed operation for <0.25 seconds

Class 4

- direct viewing is a hazard, and specular or diffuse reflections may pose a hazard
- skin protection and fire protection are concerns
- same warning label requirements as class 3B plus a locked door, door actuated power kill switch or door actuated optical filter, shutter, or equivalent
- typically >500 mW continuous output power, >10 joules per square centimeter pulsed operation

---

which can lodge in the skin. Tools for fusion splicing produce sufficient heat to melt glass fibers, and either the equipment or a freshly heated fiber splice can cause burns. Chemical safety precautions should be used when handling index matching gels, acetone, or optical cleaning chemicals. Alcohol is sometimes used for cleaning fibers and connectors, but can be flammable if used improperly. Compressed air canisters used for dusting optical components should not be punctured or shipped by air freight to avoid explosive decompression. Normal safety precautions for working with electronics equipment and high voltage or current levels may also apply. In general, only a certified, trained professional should be allowed to work unsupervised with optical fiber, optoelectronics, or similar equipment.

2.2 Light Emitting Diodes

Light emitting diodes used in data communications are solid state semiconductor devices; to understand their function, we must first describe a bit of semiconductor physics. For the interested reader, other introductory references to solid state physics, semiconductors, and condensed matter are available [1, 2]. In a solid state device, we can speak of charge carriers as either electrons or holes (the "spaces" left behind in the material when electrons are absent). The carriers are limited to occupying certain energy states; the electron potential can be described in terms of **conduction bands** and **valence bands,** rather than individual potential wells (Figure 2.1). The highest energy level containing electrons is called the **Fermi Level.** If a material is a **conductor,** the conduction and valence bands overlap and charge carriers (electrons or holes) flow freely; the material carries an electrical current. An **insulator** is a material for which there is a large enough gap between the conduction and valence bands to prohibit the flow of carriers; the Fermi level lies in the middle of the forbidden region between bands, called the **band gap.** A **semiconductor** is a material for which the band gap is small enough that carriers can be excited into the conduction band with some stimulus; the Fermi level lies at the edge of the valence band (if the majority of carriers are holes) or at the edge of the conduction band (if the majority carriers are electrons). The first case is called a **p-type** semiconductor (since the carriers are positively charged holes), the second is called **n-type** (since the carriers

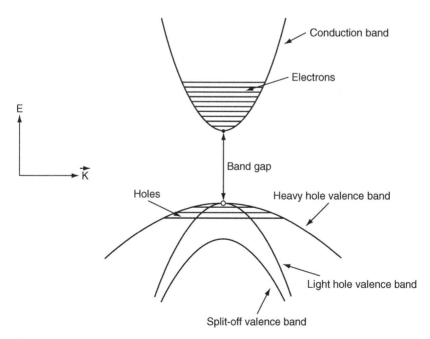

Figure 2.1 Energy band structure (energy vs. k) for direct band gap semiconductor.

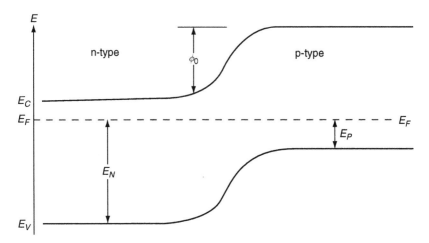

Figure 2.2   Conduction and valence bands of a semiconductor.

are negatively charged electrons). When a p-type and an n-type layer are sandwiched together, they form a PN junction (Figure 2.2). PN junctions are useful for generating light because when they are forward biased, electrons are injected from the N region and holes are injected from the P region into the active region. Free electrons and free holes can recombine across the band gap and emit photons (called **spontaneous emission**) with energy near the band gap, resulting in an LED. Like any other diode, an LED has a characteristic current–voltage response and will pass current only in one direction. Its optical output power depends on the current density, which in turn depends on the applied voltage; below a certain threshold current, the optical power is negligible.

In fiber communications systems, LEDs are used for low-cost, high reliability sources, typically operating with 62.5/125 micron graded-index multimode fiber at data rates up to a few hundred Mb/s or perhaps as high as 1 Gbit/s in some cases. For short distances, GalliumArsenide (GaAs) based LEDs operating near 850 nm are used because of their low cost and low temperature dependence. For distances up to ~10 km, more complex LEDs of InGAAsP grown on InP and emitting light at 1.3 microns wavelengths are often used [2]. The switching speed of an LED is proportional to the electron-hole recombination rate, R, given by

$$R = \frac{J}{de} \tag{2.1}$$

where J is the current density in A/m^2, d is the thickness of the recombination region, and e is the charge of an electron. The optical output power of an LED is proportional to the drive current, I, according to

$$P_{\text{out}} = I \left(\frac{Qhc}{e\lambda} \right) \qquad (2.2)$$

where Q is the quantum efficiency, h is Planck's constant, and λ is the wavelength of light.

Spontaneous emission results in light with a broad spectrum which is distributed in all directions inside the active layer. The wavelength range emitted by the LED widens as its absolute junction temperature, T, increases, according to

$$\Delta\lambda = 3.3 \left(\frac{kT}{h} \right) \left(\frac{\lambda^2}{c} \right) \qquad (2.3)$$

where k is Boltzmann's constant and c is the speed of light. As the LED temperature rises, its spectrum shifts to longer wavelengths and decreases in amplitude.

There are two types of LEDs used in communications, surface emitting LEDs and ELEDs. Thus, there are two ways to couple as much light as possible from the LED into the fiber. Butt coupling a multimode fiber to the LED surface (as shown in Figure 2.3) is the simplest method, since a surface emitting geometry allows light to come out the top (or bottom) of the semiconductor structure. The alternative is to cleave the LED, and collect the light that is emitted from the edge, thus ELEDs are formed (Figure 2.4). While surface emitting LEDs are typically less expensive, ELEDs have the advantage that they can be used with single-mode fiber. Both can be modulated up to around 622 Mb/s.

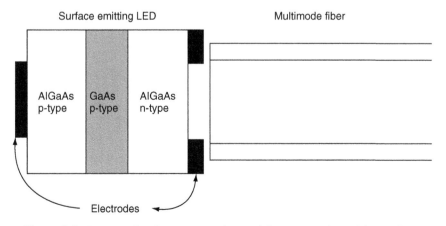

Figure 2.3   Butt-coupling between a surface emitting LED and a multimode fiber.

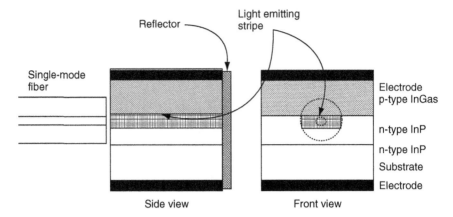

Figure 2.4 An edge emitting LED.

For a surface emitting LED, the cone of spontaneous emission that can be accepted by a typical fiber (numerical aperture NA = 0.25) is only ~0.2 percent. This coupling is made even smaller by reflection losses (such as Fresnel reflection). To improve the coupling efficiency, lenses and mirrors are used in several different geometries (e.g. see Figure 2.5). Typical operating specifications for a surface emitting LED at 1.3 microns pigtailed to a 62.5/125 micron core graded-index fiber might be $16\,\mu$W at 100 mA input current, for ~0.02 percent overall efficiency with a modulation capability of 622 Mb/s. They are usually packaged in lensed TO-18 or TO-46 cans, although SMA and ST connectors are also used.

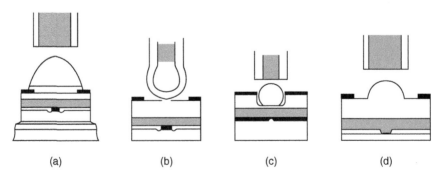

Figure 2.5 Typical geometries for coupling from LEDs into fibers: (a) hemispherical lens attached with encapsulating plastic; (b) lensed fiber tip; (c) microlens aligned through use of an etched well; and (d) spherical semiconductor surface formed on the substrate side of the LED.

Edge emitting LEDs have greater coupling efficiencies because their source area is smaller. Their geometry is similar to a conventional diode laser, because light travels back and forth in the plane of the active region and is emitted from one anti-reflection coated cleaved end. These LEDs differ from lasers in that they do not have a feedback cavity. Because of their smaller source size, they can be coupled into single mode fiber with modest efficiency ($\sim$0.04%). ELEDs are typically used as low-coherence sources for fiber sensor applications, rather than in communications, because of their broad emission spectrum.

Device characteristic curves for LEDs include the forward current vs. applied voltage, the light emission or luminous intensity vs. forward current, the spectral distribution and the luminous intensity vs. temperature and angle. An example of these is shown in Figure 2.6. Since LEDs are chosen for low cost applications, they are modulated by varying the drive current, rather than using an external modulator.

2.3 Lasers

There are many different types of lasers, only some of which are useful for communication systems. The term "**laser**" is an acronym for Light Amplification by Stimulated Emission of Radiation. In order to form a basic laser system, we require a media which can produce light (photons) through **stimulated emission**. This means that electrons in the media are excited to a higher than normal energy state, then decay into a lower energy state, releasing photons in the process. These photons will have the same wavelength, or frequency, as well as the same phase; thus, they will add up to create a **coherent** optical beam which emerges from the laser. Although we speak of laser light as being **monochromatic**, or having a single wavelength, in reality a laser produces a narrow spread of different wavelengths, or an emission spectra. The material which creates photons in this way, called the **gain media**, is most commonly a gas (in the case of medical lasers) or a semiconductor material (in the case of some medical devices and almost all communication systems). Electrons are excited by supplying energy from some external source, a process called **pumping** the laser. Most lasers are pumped with an electrical voltage, though another optical beam can sometimes be used to provide the necessary energy. Since the electrons are normally found in their lower energy state, pumping them into a higher energy state creates what is known as a **population inversion**. The final element in creating

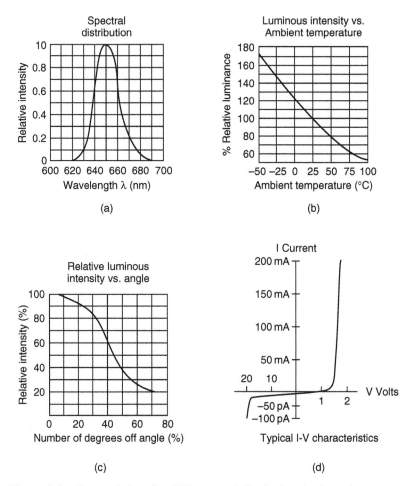

Figure 2.6 Characteristics of an LED: spectral distribution, luminous intensity vs. temperature and angle, and I vs. V characteristics.

a laser is an optical feedback mechanism, such as a pair of mirrors, to reflect light many times through the gain media and produce a stronger output beam. Laser light is coherent, meaning that all the photons are at the same wavelength (frequency) and all are in phase with each other. Lasers have very low divergence (a laser spot will spread out very little over long distances).

In most fiber optic communication systems, we are interested in the type of lasers based on solid state devices, similar to LEDs in that they use PN heterojunctions to emit photons. In LEDs, PN junctions are used for generating light because when they are forward biased, electrons are injected from the N region and holes are injected from the P region into the active region. Free electrons and free holes can recombine across the band gap and emit photons (called spontaneous emission) with energy near the band gap. Lasers are different from LEDs because the light is produced by **stimulated emission** of radiation which results in light of a very narrow, precise wavelength. As noted above, the semiconductor laser diode requires three essential operation elements:

1. Optical feedback from mirrors on either side of the laser cavity (originally provided by a cleaved facet or multilayer Fresnel reflector, more recently provided by wide band-gap layers of cladding).
2. A gain medium (the PN heterojunction, with spontaneous emission providing optical gain).
3. A pumping source (applied current).

The optical cavity used for some types of lasers is based on a **Fabry-Perot (FP)** device, commonly described in basic physics, in which the length of a mirrored cavity is an integral multiple of the laser wavelength. Compared with LED emission, laser sources exhibit higher output power (3–100 mW or more), spread over a narrower band of wavelengths (<10 nm), are more directional (beam divergence 5–10 degrees), have a greater modulation bandwidth (hundreds of MHz to several GHz or more), and are coherent (emit light at a single frequency and phase) [3].

The minimum requirement for lasing action to occur is that the light intensity after one complete trip through the cavity must at least be equal to its starting intensity, so that there is enough gain in the optical signal to overcome losses in the cavity. This **threshold condition** is given by the relationship

$$R_1 R_2 \exp(2(g - \alpha)L) = 1 \tag{2.4}$$

where the reflectivity of the cavity mirrors is R_1 and R_2, the lasing cavity is of length L, and the material has a gain of g and a loss of α.

The **threshold current** is the minimum current to produce laser light. This is a measured operating characteristic and for double heterojunction, lasers should be under $\sim 500\,\text{A/cm}^2$ at $1.3\,\mu\text{m}$ and $\sim 1000\,\text{A/cm}^2$ at $1.55\,\mu\text{m}$.

The laser emits photons whose energy, E, and wavelength, λ (or frequency, f) are related by

$$E = hf = \frac{hc}{\lambda} \tag{2.5}$$

where h is Planck's constant. For semiconductor laser diodes, photons are produced when electrons that were pumped into a higher energy state drop back into a lower energy state. The difference in energy states is called the **band gap**, Eg, and the wavelength of light produced is given by

$$\lambda(\text{microns}) = \frac{Eg\ (\text{electron volts})}{hc} = \frac{Eg}{1.24} \tag{2.6}$$

When we discuss photodetectors in a later chapter, we will see that the reverse process is also useful; incident light which illuminates a semiconductor device can create an electrical current by exciting electrons across the band gap. The band gap in many semiconductor compounds can be changed by doping the material to produce different wavelengths of light.

Given that a laser diode with an active area of width w and length L will have a current $I = JwL$, and the current density is related to the carrier density through $J = eNd/\tau$, where τ is the lifetime of the electron/hole pairs, N is the carrier density, and d is the layer thickness, the **threshold current** required to achieve laser operation can be calculated from

$$I_{\text{th}} = I_{\text{tr}} + ewN(1/2 \ln\ (1/R_1 R_2) + \alpha L)d\frac{(1 + 2/V^2)}{\tau a_{\text{L}}} \tag{2.7}$$

where I_{tr} is the transparency current, a_{L} is the proportionality constant between the gain and the carrier density near transparency and the waveguide V parameter is calculated from the requirement that a waveguide must be thinner than the cutoff value for higher order modes for proper optical confinement,

$$V = dk\ \sqrt{(n_{\text{g}}^2 - n_{\text{c}}^2)} \tag{2.8}$$

where n_{g} is the refractive index of the waveguide area, n_{c} is the refractive index of the cladding, d is the thickness of the waveguide layer and $k = 2\pi/\lambda$.

Measured curves of light output vs. current are also shown on most specifications. When operating above threshold, the output power can be calculated from

$$P = (h\upsilon/e)(\alpha/(\alpha + \alpha_{\text{i}}))(I - I_{\text{th}} - I_{\text{L}})\eta \tag{2.9}$$

where α_i is any internal losses for the laser mode, and η is the quantum efficiency. Calculating the quantum efficiency is beyond the scope of this book, but can be found in reference [4].

Multiple PN junctions can be used in a single laser device, and there are many different types of lasers which can be made in this way. Structurally, a semiconductor DH laser is similar to ELEDs, except that photons and carriers are confined by growing the active region as a thin layer and surrounding it with layers of wide band-gap material. When the active layer is as thin as a few tens of nanometers, the free electron and hole energy levels become quantized, and the active layer becomes a **quantum well**. Because of their high gain, one or more quantum wells are often used as the active layer. While the double heterostructure and quantum well lasers use cleaved facets to provide optical feedback, distributed feedback and distributed Bragg reflector lasers replace one or more facet reflectors with a waveguide grating located outside of the active region to fine-tune the operating wavelength.

Vertical cavity surface emitting lasers have significantly different architecture, which has advantages for low cost data transmission. The light is emitted directly from the surface, not from the edge, which means fiber can be directly butt-coupled. Existing commercial VCSELs emit light exclusively at short wavelengths (near 850 nm). Long wavelength VCSELs (1.3 and 1.55 μm) have been demonstrated in laboratories but are not yet commercially available – there are a number of technical issues to be addressed, including reliability, poor high temperature characteristics and low reflectivity because of their InP/InGaAsP Bragg mirrors. There are solutions to these problems under development, such as providing different types of dielectric mirrors, wafer fusion, metamorphic Bragg reflectors or wafer fusing GaAs/AsGaAs Bragg mirrors to the InP lasers.

2.3.1 DOUBLE HETEROSTRUCTURE LASER DIODES

The standard telecommunications source is an edge-emitting laser, grown with an active layer that has a band gap near either 1.5 or 1.3 microns. The layers are made of InGaAsP crystals, grown so they are lattice matched to InP for long wavelengths, or $Al_yGa_{1-y}As/Al_xGa_{1-x}As$ ($x > y > 0$) for near infrared wavelength operation. The typical geometry is shown in Figure 2.7.

The current is confined to a stripe region which is the width of the optically active area, for efficiency. While the simplest laser diode structures do not specifically confine light, except as a result of their stripe

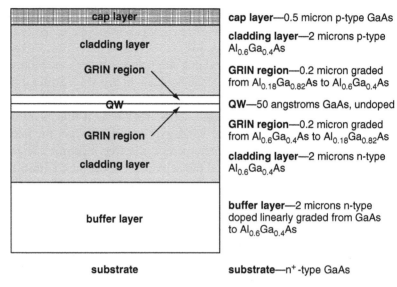

cap layer	cap layer—0.5 micron p-type GaAs
cladding layer	cladding layer—2 microns p-type $Al_{0.6}Ga_{0.4}As$
GRIN region	GRIN region—0.2 micron graded from $Al_{0.18}Ga_{0.82}As$ to $Al_{0.6}Ga_{0.4}As$
QW	QW—50 angstroms GaAs, undoped
GRIN region	GRIN region—0.2 micron graded from $Al_{0.6}Ga_{0.4}As$ to $Al_{0.18}Ga_{0.82}As$
cladding layer	cladding layer—2 microns n-type $Al_{0.6}Ga_{0.4}As$
buffer layer	buffer layer—2 microns n-type doped linearly graded from GaAs to $Al_{0.6}Ga_{0.4}As$
substrate	substrate—n^+-type GaAs

Figure 2.7 A GaAs/AlAs graded-index (GRIN) SQW laser structure.

geometry and carrier injection direction (known as **gain-guided**), these are not usually used for telecommunications applications. Higher quality lasers confine the light using buried heterostructures (BH) or ridge waveguides (RWG). The DH semiconductor laser represents the single largest constituent of today's semiconductor laser production because of its application in compact disk (CD) players and CD data storage.

2.3.2   QUANTUM WELL (QW) AND STRAINED LAYER (SL) QUANTUM WELL LASERS

When the active layer of a semiconductor laser is so thin that the confined carriers have quantized energies, it becomes an QW. The main advantage to this design is higher gain. In practice, either multiple QWs must be used, or else we require a guided wave structure that focuses the light on a single quantum well (SQW) (Figure 2.8). Strained layer (SL) quantum wells have advanced the performance of InP lasers. In a bulk semiconductor, there are two valence bands (called heavy hole and light hole) which are at the same level as the potential well minimum. QWs separate these levels, which allow population inversion to become more efficient. Strain moves these levels even further apart, making them perform even more efficiently (Figure 2.9). Typical 1.3 micron and 1.5 micron InP lasers use 5 to 15 internally strained QWs.

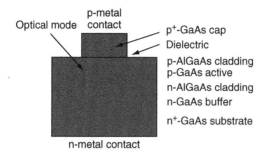

Figure 2.8 Schematic diagram of an ridge waveguide GaAs/AlGaAs double-heterojunction (DH) laser.

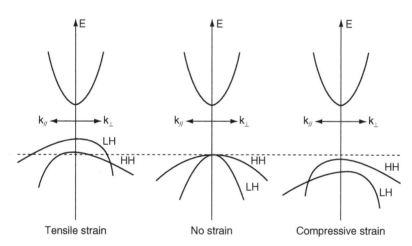

Figure 2.9 Schematic energy band diagram in k space showing the removal of degeneracy between heavy hole (HH) valence band edge and light hole valence band edge for both compressive and tensile strained InGaAs gain medium.

2.3.3 DISTRIBUTED BRAGG REFLECTOR (DBR) AND DISTRIBUTED FEEDBACK (DFB) LASERS

The DBR replaces the facet reflectors with a diffraction grating outside of the active region (see Figure 2.10). The Bragg reflector will have wavelength dependent reflectivity with resonance properties. The resulting devices behave like FP lasers, except that the narrow resonance gives them single-mode operation at all excitation levels.

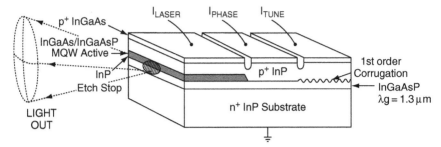

Figure 2.10 Schematic for DBR laser configuration in a geometry that includes a phase portion for phase tuning and a tunable DBR grating. Fixed-wavelength DBR lasers do not require this tuning region. Designed for 1.55 μm output, light is waveguided in the transparent layer below the MQW that has a bandgap at a wavelength of 1.3 μm. The guided wave reflects from the rear grating, sees gain in the MQW active region, and is partially emitted and partially reflected from the cleaved front facet. Fully planar integration is possible if the front cleave is replaced by another DBR grating [5].

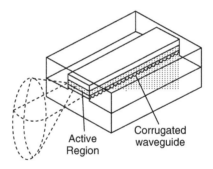

Figure 2.11 Geometry for a DFB laser, showing a buried grating waveguide that forms the separate confinement heterostructure laser, which was grown on top of a grating etched substrate. The cross-hatched region contains the MQW active layer. A stripe mesa is etched and regrown to provide a buried heterostructure laser. Reflection from the cleaved facets must be suppressed by means of an antireflection coating.

Distributed feedback lasers are similar to DBR lasers, except that the Bragg reflector is placed directly on the active region or its cladding (Figure 2.11). There is no on-resonance solution to the threshold condition in this geometry. This is resolved by either detuning to off-resonance solutions, or adding a quarter wavelength shifted grating.

Once the DFB laser is optimized, it will be single mode at essentially all power levels and under all modulation conditions. This is a significant advantage; over the past few years, many telecommunication applications have adopted strained QW-distributed feedback lasers at 1.3 micron or 1.55 micron.

2.3.4 VERTICAL CAVITY SURFACE-EMITTING LASERS

Vertical cavity surface-emitting lasers incorporate many of the advances described in the previous sections, such as QWs and DBR, into a surface emitting geometry that has many theoretical and practical advantages. While this design is difficult to achieve for long wavelengths (1.3 and 1.55 micron), it is very popular in 850-nm versions which have found wide applications in consumer electronics. These sources are low cost and compatible with inexpensive silicon detectors.

A diagram of a typical VCSEL is shown in Figure 2.12. The VCSEL has mirrors on the top and bottom of the active layer, forming a vertical cavity. DBRs (multiplayer quarter wavelength dielectric stacks) form high reflectance mirrors. A typical active region might consist of three QWs each with 10 nm thickness, separated by about 10 nm. It is a technical challenge to inject carriers from the top electrode down through the Bragg reflector, because the GaAs layers provide potential wells that trap carriers. The operating voltage is two to three times that of edge emitting lasers. The light emitted from VCSELs will have similar performance to edge emitting lasers, with a more graceful turn-on due to more spontaneous emission. More information about the operating characteristics can be found in reference [4].

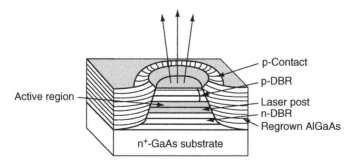

Figure 2.12 One example of an VCSEL geometry. This is a passive antiguide region (PAR) VCSEL. Light is reflected up and down through the active region by the two DBR mirrors. After the laser post is etched, regrowth in the region outside the mesa provides a high refractive index AlGaAs*nipi* region to stop current flow and to provide excess loss to higher order modes.

2.4   Modulators

In an optical communication system, an electrical signal is converted to an optical signal by **modulating** the source of light. Turning the source on and off is **direct modulation,** but often a **modulator** is used to switch a constant source of light. As noted previously, some optical sources can be directly modulated, while others require external modulation. External modulation is primarily used in telecommunication systems for more efficient operation at higher data rates. In an CW laser, the optical output is constant over time, and the laser light is produced in an uninterrupted fashion. The alternative is pulsed or modulated lasers. While LEDs may directly modulate their signal by turning on and off their electrical power, designs for much higher data rates may use CW lasers in conjunction with an external optical modulator. This is done to reduce effects such as turn-on delay, relaxation oscillation, mode hopping and/or frequency chirping.

Turn-on delay occurs in semiconductor lasers because it takes time for the carrier density to reach its threshold value and for the photon density to build up to a critical value. There is thus a small delay between application of a voltage to the laser and the resulting optical output (see Figure 2.13).

Relaxation oscillations are overshoots of the desired signal level, which occur as the photon dynamics and carrier dynamics are coming into equilibrium. This creates oscillations in the light output for short periods of time following a rise or fall in optical power (again, see Figure 2.13).

Mode hopping is the sudden shift of the laser diode output beam from one longitudinal mode to another during operation. This may cause excessive modal noise in an optical link.

Chirping occurs when the carrier density in the active region is rapidly changed, causing a shift of the index of refraction with time. This modifies the frequency of the output light (the term "chirp" refers to a linear change in frequency over time) and affects dispersion and other link noise sources. Chirping is most often observed in pulsed sources.

The **extinction ratio** of a modulator is the ratio of the optical power level generated when the light source is "on," compared to when the light source is "off." It is typically expressed as a fraction at a specific wavelength, or in dB. Put another way, for **intensity modulators** we can define the **modulation depth (MD)** in terms of the intensity when no data is transmitted, I_0, to the intensity when data is transmitted, I;

$$MD = \frac{(I_0 - I)}{I_0} \qquad (2.10)$$

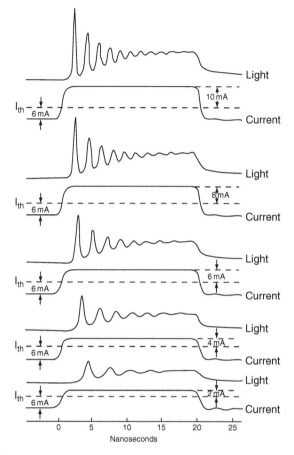

Figure 2.13 Experimental example of turn-on delay and relaxation oscillations in a laser diode when the operating current is suddenly switched from 6 mA below the threshold current of 177 mA to varying levels above threshold (from 2 to 10 mA). The GaAs laser diode was 50 μm long, with a SiO$_2$ defined stripe 20 μm wide. Light output and current pulse are shown for each case.

Then extinction ratio is the maximum value of the ratio MD(max) to MD when the intensity of the transmitted beam is at its minimum. For **phase modulators** it is defined in a similar way, provided light intensity is related to phase. For **frequency modulators** we speak instead of the **maximum frequency deviation**, $D(\text{max})$, given by

$$D(\text{max}) = \frac{|f_m - f_0|}{f_0} \qquad (2.11)$$

where f_m is the maximum frequency shift of the carrier frequency f_0. Some specification sheets define the **degree of isolation**, in dB, which is given by 10 times the log of either the extinction ratio or the maximum frequency deviation.

2.4.1 *LITHIUM NIOBATE MODULATORS*

Certain types of materials, such as lithium noibate crystals, can experience a change in their refractive index when an electric field is applied. This is known as the **electro-optic effect**. In the case of lithium niobate, choosing the crystal orientation to maximize the change in the index of refraction, this change can be calculated from

$$\Delta n_z = -n_z{}^3 r_{33} E_z \Gamma/2 = -n_z{}^3 r_{33} V\Gamma/G^2 \tag{2.12}$$

where n_z is the index of refraction along the crystal's z-axis, r_{33} is a coefficient describing the magnitude of the electro-optic effect along the new axis of the crystal (in m/V), E_z is the electric field applied in the z direction, V is the applied voltage, G is the gap between the electrodes, and Γ is a filling factor (also known as an optical-electrical field overlap parameter). The filling factor corrects for the fact that the applied field may not be uniform as it overlaps the waveguide, effectively diminishing the applied field.

The electro-optic effect produces a phase shift in the modulated light, which is given by

$$\Delta\phi = \Delta n_z kL \tag{2.13}$$

where k is $\frac{2\pi}{\lambda}$ and the electrode length is L.

Considering the index of refraction for bulk lithium niobate at a wavelength of 1.55 microns, when $G = 10$ microns and $V = 5$ volts, a phase shift of π is expected in a length $L \sim 1$ cm. Since Δn_z is directly proportional to the voltage, the electro-optic phase shift depends on the product of the voltage and the length, which leads engineers to use this as a figure of merit. This example would describe a 5 V-cm figure of merit.

Lithium niobate's electro-optic phase shift allows it be used to modulate intensity using interference in Y-branch **interferometric modulators**, as shown in Figure 2.14. The light entering the waveguide is evenly split at the Y junction, and then goes through oppositely biased lithium niobate crystals. The fraction of light transmitted is

$$\eta = [\cos(\Delta\phi/2)]^2 \tag{2.14}$$

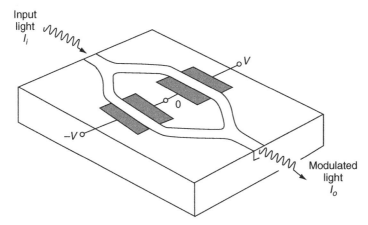

Figure 2.14 *Y*-branch interferometric modulator in the "push-pull" configuration. Center electrodes are grounded. Light is modulated by applying positive or negative voltage to the outer electrodes.

If a phase difference of π is induced, then destructive interference occurs when the light recombines, and there is no light output from the device. Otherwise there is no destructive interference, and light emerges from the device. In this way, the interferometric modulator can switch light on or off.

Modulators are used to avoid or minimize turn-on delay, relaxation oscillation, mode hopping and/or chirping. Commercially available lithium niobate modulators operate at frequencies up to 8 GHz at 1.55 microns. Direct coupling from a laser or polarization maintaining fiber is necessary, because their phase shifting design is inherently polarization dependent.

2.4.2 ELECTROABSORPTION MODULATORS

Electroabsorption refers to the electric field dependence of the absorption near the band edge of a semiconductor. This dependence is particularly strong in QWs, where it is called the quantum-confined Stark effect. There is a peak in the QW absorption spectrum known as the exciton resonance. This can form the basis for an external optical modulator.

It is desirable to build modulators from III-V semiconductors, so they can be integrated directly on the same chip as the laser, although they may also be external, coupled by means of a microlens or a fiber pigtail.

The III-V semiconductors are a good material to use in QWs, which can be designed to take advantage of electroabsorption.

When light passes through the QW material, the transmission is a function of applied voltage. The usual geometry involves several QWs in a waveguide to prevent excessive light loss from diffraction. Performance is usually characterized by two quantities, contrast ratio (CR) and insertion loss.

Contrast ratio is given by

$$CR = \frac{T_{high}}{T_{low}} = \exp(\delta\alpha L) \tag{2.15}$$

where T is transmission, and $\delta\alpha$ is the change in $\alpha(E)$, the QW absorption multiplied by the filling factor of the QWs in the waveguide, as it changes from the field E for a low signal to a high signal.

The **insertion loss** is given by

$$A = 1 - T_{high} = 1 - \exp(-\alpha_{low}L) \sim \alpha_{low}L \tag{2.16}$$

Modulator lengths are designed so the $L \sim 1/\alpha_{low}$, to keep the insertion loss low, despite the associated limitations on the contrast ratio. This results in an CR which varies as $CR = \exp(\delta\alpha/\alpha_{low})$. CR may reach as high as 10/1 or more with $<2\,\text{V}$ applied.

2.4.3   *ELECTRO OPTIC AND ELECTROREFRACTIVE SEMICONDUCTOR MODULATORS*

The electro-optic effect is not limited to lithium niobate. Some III-V semiconductors, which are easier to package because they can be integrated on the same substrate as a laser, also experience a degree of anisotropy when an electric field is applied. The advantage of this kind of system over electroabsorption is that they offer reduced chirp (frequency broadening due to time-varying refractive index).

Semiconductor modulators can be designed based on another basic property called **electrorefraction**. The change in the electric field dependence of the absorption near the band edge, electroabsorption, is closely connected to a change in the index of refraction, which can be calculated using the Kramers-Kronig relations [4]. As with electroabsorption modulators, a field is applied across a semiconductor PIN junction. The difference is that the light is refracted, rather than absorbed. This can result in a reduction in length and drive voltages required for phase modulation in waveguides.

References

[1] Burns, G. (1985) *Solid State Physics*, Chapter 10, pp. 243–349, Orlando, Fl.: Academic Press.

[2] Kittel, C. (1986) *Introduction to Solid State Physics*, Chapter 8, pp. 81–214, Wiley.

[3] Jiang, W. and Michael S. Lebby (2002) "Optical Sources: Light Emitting Diodes and Laser Technology", in *Handbook of Fiber Optic Data Communication*, C. DeCusatis (ed.), Chapter 1, pp. 41–81.

[4] Garmire, E. (2001) "Sources, Modulators and Detectors for Fiber-Optic Communications Systems", *Handbook of Optics*, M. Bass, *et al.* (ed.), pp. 4.1–4.78, McGraw-Hill.

[5] Koch, T.L. and Koran, U. (1991) *IEEE J. Quantum Electron.*, 27, p. 641.

Chapter 3 | Detectors and Receivers

In this chapter we will review optical detectors and receivers for optical communication systems. An optical receiver includes both the optical detectors, or **photodetectors** (the solid state device that convert an optical signal to an electrical signal), and various electronic circuits such as amplifiers or clock recovery circuits. We will discuss semiconductor devices such as PIN photodiodes, avalanche photodiodes (APD), photodiode arrays, Schottky-Barrier photodiodes, metal-semiconductor-metal (MSM) detectors, resonant-cavity photodetectors, and interferometric devices. APDs are used in high sensitivity applications, MSM photoconductive detectors are sometimes used because of their ease of fabrication, and Schottky-barrier photodiodes are used for high speed applications. Resonant-cavity photodetectors are not commonly used today, but have potential for future development. Other types of detectors are also becoming more common in applications such as wavelength division multiplexing (WDM) and parallel optical interconnects (POI). After a brief discussion of optical receiver circuits, we will conclude with a discussion of noise sources and their impact on the detection electronics.

Despite the fact that they both detect and measure light, optical fiber **sensors** should not be confused with fiber optic **receivers**. Unlike receivers in communication systems, fiber sensors do not directly demodulate light; rather they often use diffraction gratings to measure interference effects in systems employing optical fibers, and thereby measures environmental factors such as temperature, vibration, and strain. Often these devices require optical spectral analysis or other techniques, rather than the more simple optical power or flux measurements provided by

PIN diodes. **Interferometric sensors** are discussed briefly with other detectors.

It is worth noting that fiber optic systems typically package transmitters (discussed in Chapter 2) and receivers in a single housing known as transceivers (TRX). These are used for most two-fiber bi-directional communication links. Depending on the application, more or less electronics may be packaged within a transceiver. As before, the focus in this chapter will be mostly on communications systems; there are many types of optical detectors used in photometry, imaging systems, and other applications which are beyond the scope of this book.

Detectors are an integral part of all fiber optic communication systems. They convert a received optical signal into an electrical signal, demodulating the optical signal that was modulated at the transmitter. This initial signal may be subsequently amplified or further processed. Optical detectors must satisfy stringent requirements such as high sensitivity at the operating wavelengths, fast response speed, and minimal noise. In addition, the detector should be compact in size, use low biasing voltages and be reliable under all operating conditions. There are many different types of optical detectors available, which are sensitive over the entire optical spectra, the infrared and the ultraviolet, including such devices as photomultiplier tubes, charge coupled devices (CCDs), thermal and photoconductive detectors [1]. However, fiber optic communication systems most commonly use solid state semiconductor devices, such as the PIN photodiode.

A general semiconductor detector has basically three processes: first, carriers (electron-hole pairs) are generated by the incident light; second, carriers are transported through the detection device and the gain may be multiplied by whatever current-gain mechanism may be present; and third, the current interacts with the external circuit to provide the output signal. In this chapter, we will begin with an overview of detector terminology. This is followed by a detailed description of the types of detectors used extensively in fiber optics. The PIN photodiode is the most commonly used fiber optic detector, and almost exclusively used in transceivers.

3.1   Basic Detector Specification Terminology

Every detector specification should include a physical description, including dimensions and construction details (such as metal or plastic housing). We have tried to be inclusive in the following list of terms, which means that not all of these quantities will apply to every detector specification. Since specifications are not standardized, it is impossible to include all possible terms used; however, most detectors are described by certain standard figures of merit which will be discussed in this section. It is important to consider the manufacturer's context for all values; a detector designed for a specific application may not be appropriate for a different application even though the specification seems appropriate.

There are several figures of merit used to characterize the performance of different detectors. **Responsivity**, or Response, is the sensitivity of the detector to input flux. It is given by

$$R(\lambda) = \frac{I}{\phi(\lambda)} \tag{3.1}$$

Where I is the detector output current (in amperes) and ϕ is the incident optical power on the detector (in watts). Thus, the units of responsivity are amperes per watt (A/W). Even when the detector is not illuminated, some current will flow; this **dark current** may be subtracted from the detector output signal when determining detector performance. Dark current is the thermally generated current in a photodiode under a completely dark environment; it depends on the material, doping and structure of the photodiode. It is the lowest level of thermal noise. Dark current in photodiodes limits the sensitivity (minimum detectable power). The reduction of dark current is important for the improvement of the minimum detectable power. It is usually simply measured and then subtracted from the flux, like background noise, in most specifications. However, the dark current is temperature dependent, so care must be taken to evaluate it over the expected operating conditions. If the anticipated signal is only a fraction of the dark current, then RMS noise in the dark current may mask the signal. Responsivity is defined at a specific wavelength; the term **spectral responsivity** is used to describe the variation at different wavelengths. Responsivity vs. wavelength is often included in a specification as a graph, as well as placed in a performance chart at a specified wavelength.

Quantum efficiency (QE) is the ratio of the number of electron-hole pairs collected at the detector electrodes to the number of photons in the incident light. It depends on the material from which the detector is made, and is primarily determined by reflectivity, absorption coefficient, and carrier diffusion length considerations. As the absorption coefficient is dependent on the incident light wavelength, the quantum efficiency has a **spectral response** as well. QE is the fundamental efficiency of the diode for converting photons into electron-hole pairs. For example, the QE of a PIN diode can be calculated by

$$QE = (1 - R)T(1 - e^{-\alpha W})\qquad(3.2)$$

where R is the surface reflectivity, T is the transmission of any lossy electrode layers, W is the thickness of the absorbing layer, and α is the absorption coefficient.

Quantum efficiency affects detector performance through the responsivity (R) which can be calculated

$$R(\lambda) = \frac{QE\,\lambda q}{hc}\qquad(3.3)$$

where q is the charge of an electron (1.6×10^{-19} coulomb), λ is the wavelength of the incident photon, h is Planck's constant (6.626×10^{-34} W), and c is the velocity of light (3×10^8 m/s). If wavelength is in microns and R is responsivity flux, then the units of quantum efficiency are A/W. Responsivity is the ratio of the diode's output current to input optical power and is given in A/W. A PIN photodiode typically has a responsivity of 0.6–0.8 A/W. A responsivity of 0.8 A/W means that incident light having 50 microwatts of power results in 40 microamps of current, in other words

$$I = 50\,\mu W \times 0.8\,A/W = 40\,\mu A\qquad(3.4)$$

where I is the photodiode current. For an APD, a typical responsivity is 80 A/W. The same 50 microwatts of optical power now produces 4 mA of current;

$$I = 50\,\mu W \times 80\,A/W = 4\,mA\qquad(3.5)$$

The minimum power detectable by the photodiode determines the lowest level of incident optical power that the photodiode can detect. It is related to the **dark current** in the diode, or the current which flows when no light is present; the dark current will set the lower limit. Other noise sources will be discussed later and are also included in detector specifications, including those associated with the diode and those associated with the receiver. The **noise floor** of a photodiode, which tell us the minimum detectable power, is the ratio of noise current to responsivity.

$$\text{Noise floor} = \frac{\text{noise}}{\text{responsivity}}\qquad(3.6)$$

For initial evaluation of a photodiode, we can use the dark current to estimate the noise floor. Consider a photodiode with R = 0.8 A/W and a dark current of 2 nA. The minimum detectable power is

$$\text{Noise floor} = \frac{2\,\text{nA}}{0.8\,\text{nA/nW}} = 2.5\,\text{nW} \tag{3.7}$$

More precise estimates must include other noise sources, such as thermal and shot noise. As discussed, the noise depends on current, load resistance, temperature, and bandwidth.

Response time is the time required for the photodiode to respond to an incoming optical signal and produce an external current. Similarly to a source, response time is usually specified as a **rise time** and a **fall time**, measured between the 10 and 90% points of amplitude (other specifications may measure rise and fall times at the 20–80% points, or when the signal rises or falls to $1/e$ of its initial value.

As we have seen in prior chapters, **bandwidth**, or the difference between the highest and the lowest frequency that can be transmitted (see Section 1.2 and Section 2.1, etc.) is an essential feature of any fiber-optic communications system. The bandwidth of a photodiode can be limited by its **timing response** (its rise time and fall time) or its RC time constant, whichever results in the slower speed or bandwidth. The bandwidth of a circuit limited by the RC time constant is

$$B = \frac{1}{2\pi RC} \tag{3.8}$$

where R is the load resistance and C is the diode capacitance. Figure 3.1 shows the equivalent circuit model of a photodiode. It consists of a current source in parallel with a resistance and a capacitance. It appears as a low-pass filter, a resistor–capacitor network that passes low frequencies and attenuates high frequencies. The **cutoff frequency** is the highest frequency (or wavelength) for which the photodetector is sensitive; in practice, it is often defined as the frequency for which the signal is

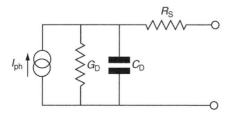

Figure 3.1   Small-signal equivalent circuit for a reversed biased photodiode.

attenuated 3 dB (this is also called the **3 dB bandwidth**). Photodiodes for high speed operation must have a very low capacitance. The capacitance in a photodiode is mainly the junction capacitance formed at the PN junction, as well as any capacitance contributed by the packaging.

Bias voltage refers to an external voltage applied to the detector, and will be more fully described in the following section. Photodiodes require bias voltages ranging from as low as 0 V for some PIN photodiodes to several hundred volts for APDs. Bias voltage significantly affects operation, since dark current, responsivity, and response time all increase with bias voltage. APDs are usually biased near their avalanche breakdown point to ensure fast response.

Active Area and **Effective Sensing Area** are fairly straightforward; they measure the size of the detecting surface of the detection element. The **uniformity** of response refers to the percentage change of the sensitivity across the active area. Operating temperature is the temperature range over which a detector is accurate and will not be damaged by being powered. However, there may be changes in sensitivity and dark current which must be taken into account for specific applications. Storage temperature will have a considerably larger range; basically, it describes the temperature range under which the detector will not melt, freeze or otherwise be damaged or lose its operating characteristics.

Noise Equivalent Power (NEP) is the amount of flux that would create a signal of the same strength as the RMS detector noise. In other words, it is a measure of the minimum detectable signal; for this reason, it is the most commonly used version of the more generic figure of merit, Noise Equivalent Detector Input. More formally, it may be defined as the optical power (of a given wavelength or spectral content) required to produce a detector current equal to RMS noise in a unit bandwidth of 1 Hz.

$$NEP(\lambda) = \frac{i_n(\lambda)}{R(\lambda)} \tag{3.9}$$

where i_n is the RMS noise current and R is the responsivity, defined above. It can be shown [2] that to a good approximation,

$$NEP = \frac{2hc}{QE\lambda} \tag{3.10}$$

where this expression gives the NEP of an ideal diode when QE = 1. If the dark current is large, this expression may be approximated by

$$NEP = \frac{hc\sqrt{2qI}}{QEq\lambda} \tag{3.11}$$

where I is the detector current. Sometimes it is easier to work with **detectivity**, which is the reciprocal of NEP. The higher the detectivity, the smaller the signal a detector can measure; this is a convenient way to characterize more sensitive detectors. Detectivity and NEP vary with the inverse of the square of active area of the detector, as well as with temperature, wavelength, modulation frequency, signal voltage, and bandwidth. For a photodiode detecting monochromatic light and dominated by dark current, detectivity is given by

$$D = \frac{QEq\lambda}{hc\sqrt{2qI}} \tag{3.12}$$

The quantity-specific detectivity accounts for the fact that dark current is often proportional to detector area, A; it is defined by

$$D^* = D\sqrt{A} \tag{3.13}$$

Normalized detectivity is detectivity multiplied by the square root of the product of active area and bandwidth; this product is usually constant, and allows comparison of different detector types independent of size and bandwidth limits. This is because most detector noise is white noise (Gaussian power spectra), and white noise power is proportional to the bandwidth of the detector electronics; thus the noise signal is proportional to the square root of bandwidth. Also, note that electrical noise power is usually proportional to detector area and the voltage which provides a measure of that noise is proportional to the square root of power. Normalized detectivity is given by:

$$D_n = D\sqrt{AB} = \frac{\sqrt{AB}}{\text{NEP}} \tag{3.14}$$

where B is the bandwidth. The units are $\sqrt{\text{cm Hz}} \ \text{W}^{-1}$. Normalized detectivity is a function of wavelength and spectral responsivity; it is often quoted as normalized spectral responsivity.

Bandwidth, B, is the range of frequencies over which a particular instrument is designed to function within specified limits (see Equation 3.8). Bandwidth is often adjusted to limit noise; in some specifications it is chosen as 1 Hz, so NEP is quoted in watts/Hz. Wide bandwidth detectors required in optical data communication often operate at a low resistance and require a minimal signal current much larger than the dark current; the load resistance, amplifier, and other noise sources can make the use of NEP, D, D^*, and D_n inappropriate for characterizing these applications.

Linear range is the range of incident radiant flux over which the signal output is a linear function of the input. The lower limit of linearity is NEP, and the upper limit is saturation. **Saturation** occurs when the detector begins to form less signal output for the same increase of input flux; increasing the input signal no longer results in an increase in the detector output. When a detector begins to saturate, it has reached the end of its linear range. **Dynamic range** can be used to describe non-linear detectors, like the human eye. Although communication systems do not typically use filters on the detector elements, neutral density filters can be used to increase the dynamic range of a detector system by creating islands of linearity, whose actual flux is determined by dividing output signals of the detector by the transmission of the filter. Without filtering, the dynamic range would be limited to the linear range of the detector, which would be less because the detector would saturate without the filter to limit the incident flux. The units of linear range are incident radiant flux or power (watts or irradiance).

Measuring the response of a detector to flux is known as **calibration**. Some detectors can be self-calibrated, others require manufacturer calibration. Calibration certificates are supplied by most manufacturers for fiber optic test instrumentation; they are dated, and have certain time limits. The gain, also known as the amplification, is the ratio of electron hole pairs generated per incident photon. Sometimes detector electronics allows the user to adjust the gain. Wiring and pin output diagrams tell the user how to operate the equipment, by schematically showing how to connect the input and output leads.

3.2   PN Photodiode

We will limit our discussion in this chapter to photodiode detectors used in data communications. These are solid state devices; to understand their function, we must first describe a bit of semiconductor physics, as done in Chapter 2. For the interested reader, other introductory references to solid state physics, semiconductors, and condensed matter are available [2]. In a solid state device, the electron potential can be described in terms of conduction bands and valence bands, rather than individual potential wells (Figure 3.2). The highest energy level containing electrons is called the Fermi level. If a material is a conductor, the conduction and valence bands overlap and charge carriers (electrons or holes) flow freely; the material carries an electrical current. An insulator is a material for which

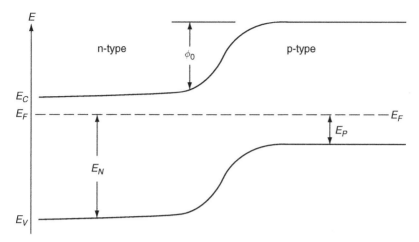

Figure 3.2 Conduction and valence bands of a semiconductor.

there is a large enough gap between the conduction and valence bands to prohibit the flow of carriers; the Fermi level lies in the middle of the forbidden region between bands, called the band gap. A semiconductor is a material for which the band gap is small enough that carriers can be excited into the conduction band with some stimulus; the Fermi level lies at the edge of the valence band (if the majority of carriers are holes) or at the edge of the conduction band (if the majority carriers are electrons). The first case is called a p-type semiconductor, the second is called n-type. These materials are useful for optical detection because incident light can excite electrons across the band gap and generate a photocurrent.

The simplest photodiode is the **PN photodiode** shown in Figure 3.3. Although this type of detector is not widely used in fiber optics, it serves the purpose of illustrating the basic ideas of semiconductor photodetection. Other devices—the PIN and APDs – are designed to overcome the limitations of the PN diode. When the PN photodiode is reverse biased (negative battery or power supply potential connected to p-type material), very little current flows. The applied electric field creates a depletion region on either side of the PN junction. Carriers—free electrons and holes—leave the junction area. In other words, electrons migrate toward the negative terminal of the device and holes toward the positive terminal. Because the depletion region has no carriers, its resistance is very high, and most of the voltage drop occurs across the junction. As a result, electrical fields are high in this region and negligible elsewhere. An incident

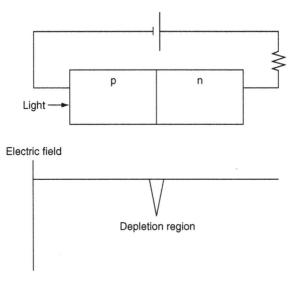

Figure 3.3 PN diode.

photon absorbed by the diode gives a bound electron sufficient energy to move from the valence band to the conduction band, creating a free electron and a hole. If this creation of carriers occurs in the depletion region, the carriers quickly separate and drift rapidly toward their respective regions. This movement sets an electron flowing as current in the external circuit. When the carriers reach the edge of the depletion region, where electrical fields are small, their movement, however, relies on diffusion mechanism. When electron-hole creation occurs outside of the depletion region, the carriers move slowly toward their respective regions. Many carriers recombine before reaching it. Their contribution to the total current is negligible. Those carriers remaining and reaching the depleted area are swiftly swept across the junction by the large electrical fields in the region to produce an external electrical current. This current, however, is delayed with respect to the absorption of the photon that created the carriers because of the initial slow movement of carriers toward their respective regions. Current, then, will continue to flow after the light is removed. This slow response, due to slow diffusion of carriers, is called slow tail response.

Two characteristics of the PN diode make it unsuitable for most fiber optic applications. First, because the depletion area is a relatively small portion of the diode's total volume, many of the absorbed photons do not contribute to the external current. The created free electrons and holes recombine before they contribute significantly to the external current. Second, the slow tail response from slow diffusion makes the diode too slow for high speed applications. This slow response limits operations to the kHz range.

3.3   PIN Photodiode

The structure of the PIN diode is designed to overcome the deficiencies of its PN counterpart. The **PIN diode** is a photoconductive device formed from a sandwich of three layers of crystal, each layer with different band structures caused by adding impurities (doping) to the base material, usually indium gallium arsenide, silicon, or germanium. The layers are doped in this arrangement: p-type (or positive) on top, intrinsic, meaning undoped, in a thin middle layer, and n-type (or negative) on the bottom. For a silicon crystal, a typical p-type impurity would be boron, and indium would be a p-type impurity for germanium [2–6]. Actually, the intrinsic layer may also be lightly doped, although not enough to make it either p-type or n-type. The change in potential at the interface has the effect of influencing the direction of current flow, creating a diode; obviously, the name PIN diode comes from the sandwich of p-type, intrinsic, and n-type layers.

The structure of a typical PIN photodiode is shown in Figure 3.4. The p-type and n-type silicon form a potential at the intrinsic region; this potential gradient depletes the junction region of charge carriers, both electrons and holes, and results in the conduction band bending. The intrinsic region has no free carriers, and thus exhibits high resistance. The junction drives holes into the p-type material and electrons into the n-type material. The difference in potential of the two materials determines the energy an electron must have to flow through the junction. When photons fall on the active area of the device they generate carriers near the junction, resulting in a voltage difference between the p-type and the n-type regions. If the diode is connected to external circuitry, a current will flow which is proportional to the illumination. The PIN diode structure addresses the main problem with PN diodes, namely providing a large depletion region for the absorption of photons. There is a tradeoff involved in the design of PIN diodes. Since most of the photons are absorbed in the intrinsic region, a thick intrinsic layer is desirable to improve photon-carrier conversion efficiency (to increase the probability of a photon being absorbed in the intrinsic region). On the other hand, a

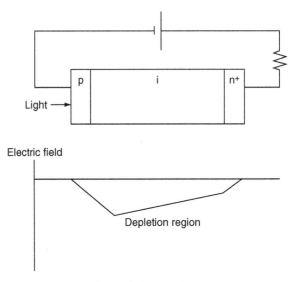

Figure 3.4 PIN diode.

thin intrinsic region is desirable for high speed devices, since it reduces the transit time of photogenerated carriers. These two conditions must be balanced in the design of PIN diodes.

Photodiodes can be operated either with or without a bias voltage. Unbiased operation is called the photovoltaic mode; certain types of noise, including $1/f$ noise, are lower and the NEP is better at low frequencies. Signal-to-noise ratio (SNR) is superior to the biased mode of operation for frequencies below about 100 kHz [6]. Biasing (connecting a voltage potential to the two sides of the junction) will sweep carriers out of the junction region faster and change the energy requirement for carrier generation to a limited extent. Biased operation (photoconductive mode) can be either forward or reverse bias. Reverse bias of the junction (positive potential connected to the n-side and negative connected to the p-side) reduces junction capacitance and improves response time; for this reason it is the preferred operation mode for pulsed detectors. A PIN diode used for photodetection may also be forward biased (the positive potential connected to the p-side and the negative to the n-side of the junction) to make the potential scaled for current to flow less, or in other words to increase the sensitivity of the detector (Figure 3.5).

An advantage of the PIN structure is that the operating wavelength and voltage, diode capacitance, and frequency response may all be predetermined during the manufacturing process. For a diode whose intrinsic

(a) Equilibrium

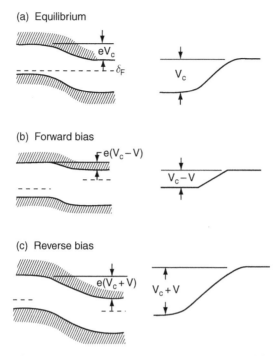

(b) Forward bias

(c) Reverse bias

Figure 3.5   Forward and reverse bias of a diode [2].

layer thickness is w with an applied bias voltage of V, the self-capacitance of the diode, C, approaches that of a parallel plate capacitor,

$$C = \frac{\varepsilon_o \varepsilon_1 A_o}{w} \tag{3.15}$$

where A_o is the junction area, ε_o the free space permittivity (8.849×10^{-12} farad/m), and ε_1 the relative permittivity. Taking typical values of $\varepsilon_1 = 12$, $w = 50$ microns, and $A_o = 10^{-7}$ m^2, $C = 0.2$ pF. QEs of 0.8 or higher can be achieved at wavelengths of 0.8–0.9 microns, with dark currents less than 1 nA at room temperature. The sensitivity of a PIN diode can vary widely by quality of manufacture. A typical PIN diode size ranges from 5 mm × 5 mm to 25 mm × 25 mm. Ideally, the detection surface will be uniformly sensitive (at the National Institute for Standards and Technology (NIST) there is a detector profiler which, by using extremely well focused light sources, can determine the sensitivity of a detector's surface [3]). For most applications, it is required that the detector is uniformly illuminated or overfilled. The spectral responsivity of an uncorrected silicon photodiode is shown in Figure 3.6. The typical

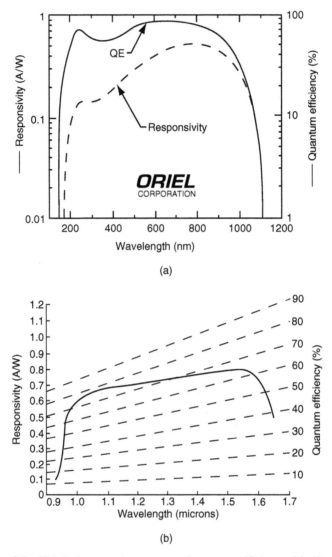

Figure 3.6 PIN diode spectral response and quantum efficiency: (a) silicon and (b) InGaAs.

QE curve is also shown for comparison. An ideal silicon detector would have zero responsivity and QE for photons whose energies are less than the band gap, or wavelengths much longer than about 1.1 micron. Just below the long wave limit, this ideal diode would have 100% QE and responsivity close to 1 A/W; responsivity vs. wavelength would be expected to follow the intrinsic spectral response of the material. In practice, this does not happen; these detectors are less sensitive in the blue region, which can sometimes be enhanced by clever doping, but not more than an order of magnitude. This lack of sensitivity is because there are fewer short wavelength photons per watt, so responsivity in terms of power drops off, and because more energetic blue photons may not be absorbed in the junction region. For color-sensitive applications such as photometry, filters are used so detectors will respond photometrically, or to the standardized CIE color coordinates; however, the lack of overall sensitivity in the blue region can potentially create noise problems when measuring a low intensity blue signal. In the deep ultraviolet, photons are often absorbed before they reach the sensitive region by detector windows or surface coatings on the semiconductor. The departure from 100% QE in real devices is typically due to Fresnel reflections from the detector surface. The long wavelength cutoff is more gradual than expected for an ideal device because the absorption coefficient decreases at long wavelengths, so more photons pass through the photosensitive layers and do not contribute to the QE. As a result, QE tends to roll off gradually near the band gap limit. This response is typical for silicon devices, which make excellent detectors in the wavelength range 0.8–0.9 microns. A common material for fiber optic applications is InGaAs, which is most sensitive in the near infrared (0.8–1.7 microns). Other PIN diode materials include HgCdZnTe for wavelengths of 2–12 microns. In the 1940s, a popular photoconductive material, lead sulfide (PbS), for infrared solid state detectors was introduced. This material is still commonly used in the region from 1 to 4 microns [6].

PIN diode detectors are not very sensitive to temperature ($-25\,^\circ\text{C} - +80\,^\circ\text{C}$) or shock and vibration, making them an ideal choice for a data communications transceiver. It is very important to keep the surface of any detector clean. This becomes an issue with PIN diode detectors because they are sufficiently rugged than they can be brought into applications which expose them to contamination. Both transceivers and optical connectors should be cleaned regularly during use to avoid dust and dirt buildup.

A sample specification for a photodiode is given in Table 3.1. Bias voltage (V) is the voltage applied to a silicon photodiode to change the potential photoelectrons must scale to become part of the signal. Bias voltage is basically a set operating characteristic in a pre-packaged detector, in this case -24 V. Shunt Resistance is the resistance of a silicon photodiode when not biased. Junction Capacitance is the capacitance of

Table 3.1 Sample specifications for a PIN diode

Receiver section

Parameter	Symbol	Test conditions	Min.	Typical	Max.	Units
Data rate (NRZ)	B	—	10	—	156	Mb/s
Sensitivity (avg)	P_{DH}	62.5 μm fiber 0.275 NA, BER $\leq 10^{-10}$	−32.5	—	−14.0	dBm
Optical wavelength	λ_{th}	—	1270	—	1380	nm
Duty cycle	—	—	25	50	75	%
Output risetime	t_{TLH}	20–80% 50Ω to V_{CC} − 2V	0.5	—	2.5	ns
Output falltime	t_{THL}	80–20% 50Ω to V_{CC} − 2V	0.5	—	2.5	ns
Output voltage	V_{DL} V_{DH}	—	V_{CC} − 1.025 V_{CC} − 1.81	—	V_{CC} − 0.88 V_{CC} − 1.62	V V
Signal detect	V_{A} V_{D}	$P_{IN} > P_{A}$ $P_{IN} < P_{D}$	V_{CC} − 1.025 V_{CC} − 1.81	—	V_{CC} − 0.88 V_{CC} − 1.62	V V
P_{IN} power levels:						
Deassert	P_{B}	—	−39.0 or P_{B}	—	−32.5	dBm
Assert	P_{A}	—	−38.0	—	−30.0	dBm
Hysteresis	—	—	1.5	2.0	—	dB
Signal detect delay time:						
Deassert	—	—	—	—	50	μs
Assert	—	—	—	—	50	μs
Power supply voltage	V_{CC} − V_{EE}	—	4.75	5.0	5.25	V
Power supply current	I_{CC} or I_{EE}	—	—	—	150	mA
Operating temperature	T_{A}	—	0	—	70	°C

Absolute maximum ratings: transceiver

Parameter	Symbol	Test conditions	Min.	Typical	Max.	Units
Storage temperature	—	—	−40	—	100	°C
Lead soldering limits	—	—	—	—	240/10	°C/s
Supply voltage	V_{CC} − V_{EE}	—	−0.2	—	7.00	V

a silicon photodiode when not biased. Breakdown Voltage is the voltage applied as a bias which is large enough to create signal on its own. When this happens, the contribution of photoelectrons is minimal, so the detector cannot function. However, once the incorrect bias is removed the detector should return to normal.

To increase response speed and QE, a variation on the PIN diode, known as the Schottky barrier diode, can be used. This approach will be treated in a later Section (3.4.3). For wavelengths longer than about 0.8 microns, a heterojunction diode may be used for this same reason. Heterojunction diodes retain the PIN sandwich structure, but the surface layer is doped to have a wider band gap and thus reduced absorption. In this case, absorption is strongest in the narrower band-gap region at the heterojunction, where the electric field is maximum; hence, good QEs can be obtained. The most common material systems for heterojunction diodes are InGaAsP on an InP substrate, or GaAlAsSb on a GaSb substrate.

A typical circuit for biased operation of a photodiode is shown in Figure 3.7, where C_a and R_a represent the input impedance of a post-amplifier. The photodiode impulse response will differ from an ideal square wave for several reasons, including transit time resulting from the drift of carriers across the depletion layer, delay caused by the diffusion of carriers generated outside the depletion layer, and the RC time constant of the diode and its load. If we return to the photodiode equivalent circuit of Figure 3.1 and insert it into this typical bias circuit, we can make the

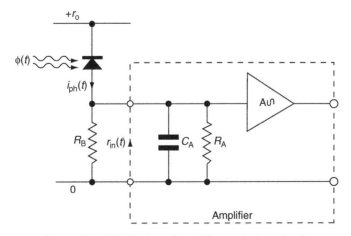

Figure 3.7   PIN diode and amplifier equivalent circuit.

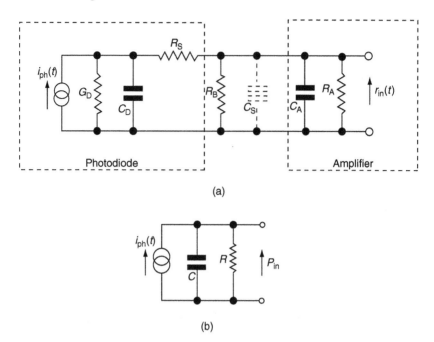

(a)

(b)

Figure 3.8 Small-signal equivalent circuit for a normally biased photodiode and amplifier: (a) complete circuit and (b) reduced circuit obtained by neglecting R_s and lumping together the parallel components.

approximation that R_s is much smaller than R_a to arrive at the equivalent small signal circuit model of Figure 3.8. In this model the resistance R is approximately equal to R_a, and the capacitance C is the sum of the diode capacitance, amplifier capacitance, and some distributed stray capacitance. In response to an optical pulse falling on the detector, the load voltage V_{in} will rise and fall exponentially with a time constant RC. In response to a photocurrent I_p which varies sinusoidally at the angular frequency

$$\omega = 2\pi f \tag{3.16}$$

the response of the load voltage will be given by

$$\frac{V_{in}(f)}{I_p(f)} = \frac{R}{1 + j2\pi f CR} \tag{3.17}$$

To obtain good high frequency response, C must be kept as low as possible; as discussed earlier, the photodiode contribution can normally be kept well below 1 pF. There is ongoing debate concerning the best approach to improve high frequency response; we must either reduce R or provide high frequency equalization. As a rule of thumb, no equalization is needed if

$$R < 1/2\pi C \Delta f \qquad (3.18)$$

where Δf is the frequency bandwidth of interest. We can also avoid the need for equalization by using a transimpedance feedback amplifier, which is often employed in commercial optical datacom receivers. These apparently simple receiver circuits can exhibit very complex behavior, and it is not always intuitive how to design the optimal detector circuit for a given application. A detailed analysis of receiver response, including the relative noise contributions and tradeoffs between different types of photodiodes, is beyond the scope of this chapter; the interested reader is referred to several good references on this subject [4–12].

3.4 Other Detectors

The PIN photodiode is definitely the most commonly used type of detector in the fiber optic communication industry. But there are a number of detectors that offer advantages in more complex situations, or for specialized types of fiber signaling; we will discuss them in the following sections. For example, POIs require detector arrays, which can be formed from either PIN diodes or MSM photoreceivers. In WDM, the optical signal traveling over one fiber optic cable is wavelength-separated, coded, and recombined. To properly receive and interpret this combined signal there is a need for more sensitive detectors, such as APDs, as well as position-sensitive detectors, such as photodiode arrays and MSM photoreceiver arrays, and also wavelength-sensitive detectors, such resonant-cavity enhanced photodetectors. The rising importance of WDM hints at the importance of growing data rates. Schottky-barrier photodiodes have always been considered for situations which required faster response than PIN diodes (100 Ghz modulation has been reported), but less sensitivity. MSM detectors, which is a Schottky diode based planar detectors, are cheap and easy to fabricate, making them desirable for some low cost applications where there are a number of parallel channels and dense integration. Resonant-cavity enhancement (RECAP) can increase the signal from Schottky-barrier and MSM detectors, as well as making high speed, thin i layer, PIN diode detectors feasible.

3.4.1 AVALANCHE PHOTODIODE

PIN diodes can be used to detect light because when photon flux irradiates the junction, the light creates electron hole pairs with their energy determined by the wavelength of the light. Current will flow if the energy is sufficient to scale the potential created by the PIN junction. This is known as the photovoltaic effect [2, 4–6]. It should be mentioned that the material from which the top layers of a PIN diode is constructed must be transparent (and clean!) to allow the passage of light to the junction. If the bias voltage is increased significantly, the photogenerated carriers have enough energy to start an avalanche process, knocking more electrons free from the lattice which contribute to amplification of the signal. This is known as an **avalanche photodiode (APD)**; it provides higher responsivity, especially in the near infrared, but also produces higher noise due to the electron avalanche process. For a PIN photodiode, each absorbed photon ideally creates one electron-hole pair, which, in turn, sets one electron flowing in the external circuit. In this sense, we can loosely compare it to an LED. There is basically a one-to-one relationship between photons and carriers and current. Extending this comparison allows us to say that an APD resembles a laser, where the relationship is not one-to-one. In a laser, a few primary carriers result in many emitted photons. In an APD, a few incident photons result in many carriers and appreciable external current.

The structure of the APD, shown in Figure 3.9, creates a very strong electrical field in a portion of the depletion region. Primary carriers—the free electrons and holes created by absorbed photons—within this field are accelerated by the field, thereby gaining several electron volts of kinetic energy. A collision of these fast carriers with neutral atoms causes the accelerated carrier to use some of its energy to raise a bound electron from the valence band to the conduction band. A free electron-hole pair is thus created. Carriers created in this way, through collision with a primary carrier, are called secondary carriers. This process of creating secondary carriers is known as collision ionization. A primary carrier can create several new secondary carriers, and secondary carriers themselves can accelerate and create new carriers. The whole process is called **photomultiplication**, which is a form of gain process. The number of electrons set flowing in the external circuit by each absorbed photon depends on the APD's multiplication factor. Typical multiplication ranges in tens and hundreds. A multiplication factor of 50 means that, on the average, 50 external electrons flow for each photon. The phrase "on the average" is important. The multiplication factor is an average, a statistical

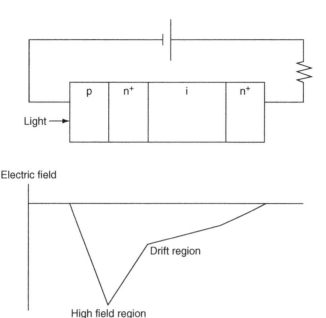

Figure 3.9 Avalanche photodiode.

mean. Each primary carrier created by a photon may create more or less secondary carriers and therefore external current. For an APD with a multiplication factor of 50, for example, any given primary carrier may actually create 44 secondary carriers or 53 secondary carriers. This variation is one source of noise that limits the sensitivity of a receiver using an APD.

The multiplication factor M_{DC} varies with the bias voltage as described in Equation (3.19)

$$M_{DC} = \frac{1}{(1 - V/V_B)^n} \tag{3.19}$$

where V_B is the breakdown voltage, and n varies between 3 and 6, depending on the semiconductor.

The photocurrent is multiplied by the multiplication factor

$$I = \frac{MQEq\phi(\lambda)\lambda}{hc} \tag{3.20}$$

as is the responsivity

$$R(\lambda) = \frac{MQE\lambda q}{hc} \tag{3.21}$$

The shot noise in an APD is that of a PIN diode multiplied by M times an excess noise factor, denoted by the square root of F, where

$$F(M) = \beta M + (1 - \beta)(2 - \frac{1}{M}) \qquad (3.22)$$

In this case, β is the ratio of the ionization coefficient of the holes divided by the ionization coefficient of the electrons. In III-V semiconductors $F = M$. It is also important to remember that the dark current of APDs is also multiplied, according to the same equations as the shot noise.

Because the accelerating forces must be strong enough to impart energies to the carriers, high bias voltages (several hundred volts in many cases) are required to create the high-field region. At lower voltages, the APD operates like a PIN diode and exhibits no internal gain. The avalanche breakdown voltage of an APD is the voltage at which collision ionization begins. An APD biased above the breakdown point will produce current in the absence of optical power. The voltage itself is sufficient to create carriers and cause collision ionization. The APD is often biased just below the breakdown point, so any optical power will create a fast response and strong output. The tradeoffs are that dark current (the current resulting from generation of electron-hole pairs even in the absence of absorbed photons) increases with bias voltage, and a high-voltage power supply is needed. Additionally, as one might expect, the avalanche breakdown process is temperature sensitive, and most APDs will require temperature compensation in datacom applications. For these reasons, APDs are not as commonly used in datacom applications as PIN diodes, despite the potentially greater sensitivity of the APD. However, they are currently being used in applications where sensitivity is very important, such as WDM.

3.4.2 PHOTODIODE ARRAY

Photodiode arrays are beginning to find applications as detectors in parallel optics for supercomputers and enhanced backplane or clustering interconnects for telecommunication products. They are also used for spectrometers (along with CCD arrays), which makes them useful to test fiber optic systems. Photodiode arrays are also being considered for WDM applications.

In the previous sections (Sections 3.2 and 3.3), we have discussed the physics guiding photodiode operation. In a photodiode array, the individual diode elements respond to incident flux by producing photocurrents, which charge individual storage capacitors. InGaAs photodiode arrays are the materials used in spectroscopic applications. Cross-talk, or signal leakage between neighboring pixels, is normally a concern in array systems. In InGaAs photodiode arrays it is limited to nearest-neighbor interactions. However, the erbium doped fiber amplifiers (EDFA) do exhibit cross-talk, so there is some discussion of using different amplifiers, such as trans-impedance amplifiers (TIA arrays). It is expected that this type of detector will be able to operate at speeds up to 10 Gbit/s (Figure 3.10) [13].

While some WDMs filter out the wavelength dependent signal, another common way to separate the signal involves using a grating to place different wavelengths at different physical positions, as is done in spectrometers. In ultra-dense WDM, as many as 100 optical channels can be used. Photodiode arrays are an obvious detector choice for these applications. In addition to detecting the signal, they offer performance feedback to the tunable lasers. At this time, photodiode arrays have not been commercially used in WDM, but their use would be similar to that in spectrometer design.

Figure 3.10 An optical demultiplexing receiver array, where data channels are focused onto an array of high-speed InGaAs photodiodes. The signals are then amplified using a trans-impedance amplifier (TIA) array.

3.4.3 SCHOTTKY-BARRIER PHOTODIODES

A variation on the PIN diode structure is shown in Figure 3.11. This is known as a **Schottky-barrier diode**; the top layer of semiconductor material has been eliminated in favor of a reverse biased, **MSM** contact. The metal layer must be thin enough to be transparent to incident light, about 10 nm; alternate structures using interdigital metal transducers are also possible. The advantage of this approach is improved QE, because there is no recombination of carriers in the surface layer before they can diffuse to either the ohmic contacts or the depletion region.

In order to make a PIN detector respond faster, we need to make the inductor layer thinner. This decreases the signal. Some architectures have been designed to try to get around this problem, by using resonant cavities (see Section 3.4.5), or by placing a number of PIN junctions

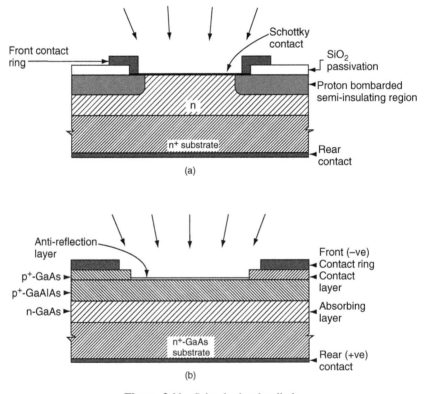

Figure 3.11   Schottky-barrier diode.

into an optical waveguide, known as traveling wave photodetectors. But Schottky-Barrier photodiodes have been another traditional alternative.

Schottky-Barrier photodiodes are not PIN based, unlike the prior detectors discussed. Instead, a thin metal layer replaces on half of the PN junction. However, it does result in the same voltage characteristics— when an incident photon hits the metal layer, carriers are created in the depletion region, and their movement again sets current flowing. The similarity in the band structure can be seen in comparing the band structure of a forward biased PIN diode in Figure 3.5 to the band structure of a forward biased n-type Schottky-Barrier diode in Figure 3.12. The voltage and field relations derived for PIN junctions can be applied to Schottky-Barrier diodes by treating the metal layer as if it were an extremely heavily doped semiconductor.

The metal layer is a good conductor, and so electrons leave the junction immediately. This results in faster operation—up to 100 Ghz modulation has been reported [14]. It also lowers the chance of recombination, increasing efficiency, and increases the types of semiconductors which can be used, since you are less limited by lattice-matching fabrication constraints. However, their sensitivity is lower than PIN diodes with the usual sized intrinsic layer. Resonant enhancement (see Section 3.4.5.) has been experimented with to increase sensitivity.

3.4.4 METAL-SEMICONDUCTOR-METAL (MSM) DETECTORS

A **MSM detector** is made by forming two Schottky diodes on top of a semi-conductor layer. The top metal layer has two contact pads and a series of "interdigitated fingers" (Figure 3.13), which form the active area of the device. One diode gets forward biased, and the other diode gets reverse biased. When illuminated, these detectors create a time-varying electrical signal. MSM photoreceiver arrays have been used as POIs. They are also being considered for use in smart pixels.

Like Schottky diodes, speed is an advantage of MSM detectors. The response speed of MSM detectors can be increased by reducing the finger spacing, and thus the carrier transit times. However, for very small finger spacing, the thickness of the absorption layer becomes comparable to the finger spacing and limits the high speed performance.

The other advantages of MSM detectors are their ease in fabrication and integration into electrical systems. That is why they are being placed in systems like arrays and smart pixels.

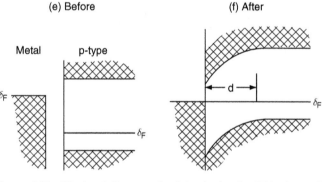

Figure 3.12 The band diagram of a Schottky-barrier. This shows the
metal-semiconductor ohmic contact with an n-type semiconductor (a) before contact
(b) after contact (c) forward bias (d) reverse bias (e) and (f) are the same
as (a) and (b), but the semiconductor is p-type [2].

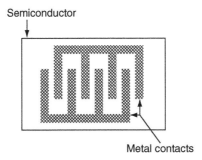

Semiconductor

Metal contacts

Figure 3.13 A diagram of an MSM detector.

MSM detectors have the same disadvantages of other Schottky based systems—weak signal and noise. They have some advantages in systems with a number of parallel channels and dense integration of detectors, which could eventually be applied to WDM systems.

As with Schottky-Barrier photodiodes, experiments have been done to increase their signal using resonant enhancement (see Section 3.4.5).

For a MSM detector deposited on a photoconductive semiconductor with a distance L between the electrodes, the photocurrent and time-varying resistance can be calculated as follows. Let us assume the detector is irradiated by a input optical flux P, at a photon energy hv.

$$Iph = \frac{GPQEq\phi(\lambda)\lambda}{hc} \tag{3.23}$$

where QE is the quantum efficiency, and G is the photoconductive gain, which is the ratio of the carrier lifetime to the carrier transit time. It can be seen that increasing the carrier lifetime decreases the speed, but increases the sensitivity, which is what you would expect.

The time varying resistance $R(t)$, is dependent on the photo induced carrier density $N(t)$.

$$N(t) = \frac{QE\phi(\lambda)\lambda}{hc} \tag{3.24}$$

$$R(t) = L/(eN(t)\mu wd) \tag{3.25}$$

Where μ is the sum of the electron and whole mobilites, w is the length along the electrodes excited by the light, and d is the effective absorption depth into the semiconductor. The specifications for a typical MSM detector are shown in Table 3.2.

Table 3.2 **Sample specifications for an MSM detector**

Detector type	MSM (metal-semiconductor-metal)
Active material	InGaAs
Bandwidth	(−3 dB electrical) 20 GHz resp. 35 GHz
Rise time	(10–90%) <11 ps
Pulse width (FWHM)	<18 ps
Wavelength range	400 nm–1.6 μm
Responsivity	0.21 A/W at 670 nm, 0.24 A/W at 810 nm, 0.19 A/W at 1.5 μm
Bias voltage	2–10V
Bias input connector	SMC male
RF signal ouput connector	K-type female
Optical input connector	FC/PC on 9 μm single-mode fiber
Maximum optical peak input power	200 mW at 20 ps, 1.3 μm
Maximum optical average input power	2 mW at 1.3 μm
Options	
Bias input connector	MSM battery case with cable
RF signal output connector	K-type male
Optical input connector on request	SMA, SC, ST, etc.
Optical input	50 μm multimode fiber, free space with collimation optics

3.4.5   *RESONANT-CAVITY ENHANCED PHOTODETECTORS (RECAP)*

The principle of **RECAP** involves placing a fast photodetector into an FP cavity to enhance the signal magnitude (Figure 3.14). In practice, the FP cavity's end mirrors are made of quarter-wave stacks of GaAs/AlAs, InGaAs/InAlAs or any other semiconductor or insulator materials which have the desired refractive index contrast. A typical example of a fast photodetector is a PIN photodetector with a very thin i-layer. By making the inductor layer thinner, you increase the speed of the detector, but decrease the signal. Schottky-Barrier photodiodes (see Section 3.4.3), and MSM detectors (see Section 3.4.4) can also have their signal magnitude increased using RECAP techniques.

By placing the detector in an FP cavity, you pass the light through it several times, which results in an increase of signal. However, due to the resonance properties of the FP cavity, this detector is highly wavelength selective. It is also highly sensitive to the position of the detector in the FP cavity, since it is a resonance effect.

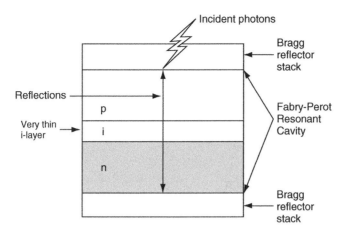

Figure 3.14   A diagram of a Resonant Cavity Photodiode.

The maximum QE, QE_{max}, of a RECAP detector can be calculated by

$$QE_{max} = \{(1 + R_2 e^{-\alpha d})(1 - R_1)/(1 - R_1 R_2 e^{-\alpha d})^2\}^*(1 - e^{-\alpha d}) \quad (3.26)$$

where R_1 and R_2 are the reflectivity of the top and bottom mirrors, α is the absorption of the active area of the detector, and d is thickness of the active area of the detector. If the reflection of the bottom mirror $R_2 = 0$, this reduces to the QE of a photodiode, only with an ideal total transmission through the top electrode (see Equation 3.2).

$$QE = (1 - R)T(1 - e^{-\alpha W}) \quad (3.2)$$

The resonant cavity enhancement, RE, can allow RCE detectors to have quantum efficiencies of nearly unity at their peak wavelength.

$$RE = \{(1 + R_2 e^{-\alpha d})/(1 - \sqrt{(R_1 R_2)}\, e^{-\alpha d})^2\} \quad (3.27)$$

The theoretical wavelength dependence of the quantum efficiency is shown in Figure 3.15.

Since WDM applications are by their very nature wavelength dependent, the wavelength selectivity of RECAP may turn out to be a useful design feature. RECAP is not yet being used in the field, but reported experimental quantum efficiencies have reached 82% for PIN diodes, and 50% improvements of photocurrent for Schottky-Barrier photodiodes [15].

3.4.6 INTERFEROMETRIC SENSORS

Optical fiber sensors are different from communications circuits in that they measure environmental factors rather than transmitting signals.

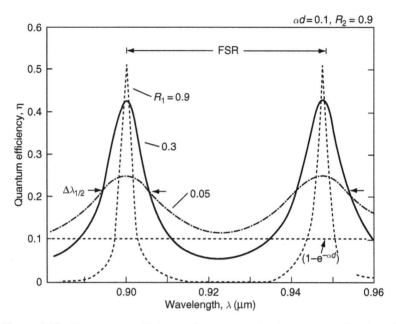

Figure 3.15   The quantum efficiency of resonant cavity detectors as a function of wavelength [15].

In Section 1.3, we discussed the most commonly used optical sensor, which is used for sensing strain and temperature in remote locations such as oil wells. FP interferometric sensors are another sensor technology. We have chosen to treat them in this chapter because of their similarity to detectors, although they could also be classified as a type of specialty fiber.

In **FP interferometric sensors** a single-mode laser diode is attached to a coupler, where one side illuminates an FP cavity and the other side a reference, index matching gel. The FP cavity is formed between an input single-mode fiber and a reflecting target element that may be a fiber (Figure 3.16). The interference between the reflected signals from the FP cavity and the reference allows us to measure strain and displacement in environments where temperature is not anticipated to change.

Fabry-Perot interferometric sensors can be either extrinsic or intrinsic. The difference between the two lies in the design of the FP cavity. Extrinsic sensors have the cavity around the fiber, and intrinsic sensors form the cavity within the optical fiber (Figure 3.17). Interferometric sensors have a high sensitivity and bandwidth, but are limited by nonlinearity in their output signals.

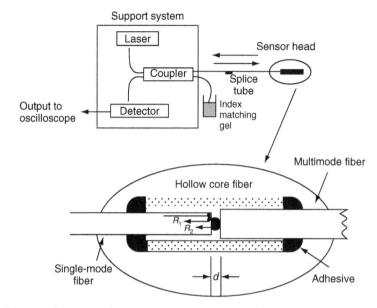

Figure 3.16   Extrinsic Fabry-Perot interferometric (EFPI) sensor and system.

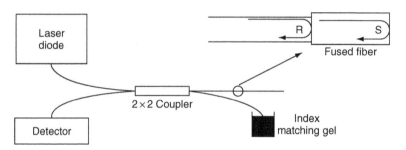

Figure 3.17   The intrinsic Fabry-Perot interferometric (IFPI) sensor.

3.5   Noise

Any optical detection or communication system is subject to various types of noise. There can be noise in the signal, noise created by the detector, and noise in the electronics. A complete discussion of noise sources has already filled several good reference books; since this is a chapter on detectors, we will briefly discuss the noise created by detectors. For a more complete discussion, the reader is referred to treatments by Dereniak and Crowe [6], who have categorized the major noise sources. The purpose of the detector is to create an electrical current in response to incident photons. It must accept highly attenuated optical energy and produce an electrical current. This current is usually feeble because of the low levels of optical power involved, often only in the order of nanowatts. Subsequent stages of the receiver amplify and possibly reshape the signal from the detector. Noise is an serious problem that limits the detector's performance. Broadly speaking, noise is any electrical or optical energy apart from the signal itself. Although noise can and does occur in every part of a communication system, it is of greatest concern in the receiver input. The reason is that receiver works with very weak signals that have been attenuated during transmission. Although very small compared to the signal levels in most circuits, the noise level is significant in relation to the weak detected signals. The same noise level in a transmitter is usually insignificant because signal levels are very strong in comparison. Indeed, the very limit of the diode's sensitivity is the noise. An optical signal that is too weak cannot be distinguished from the noise. To detect such a signal, we must either reduce the noise level or increase the power level of the signal. In the following sections, we will describe in detail several different noise sources; in practice, it is often assumed that the noise in a detection system has a constant frequency spectrum over the measurement range of interest; this is the so-called "white noise" or "Gaussian noise," and is often a combination of the effects we will describe here.

There are two kinds of amplifiers discussed in fiber optics. The first is **electronic amplifiers** that amplify the detector signal, which we will treat here. The second is **optical amplifiers**, which like repeaters and optical repeaters are placed standalone in the link to amplify the transmitter signal and extend networks to very long distances. Optical amplifiers are discussed in a later chapter.

Earlier in this chapter, we considered several circuit diagrams of detectors, including Figure 3.8, which has a PIN diode and amplifier equivalent circuit. Since we do not have all-optical communications systems, it is necessary to consider the electronic aspects of the signal-recovery process at the receiver.

Fiber optic systems typically send digital information, by which we mean a stream of binary data is transmitted by modulating the optical source such that the energy emitted during each bit period is at one of two levels. The high corresponds to 1 and the low corresponds to 0, even though the low is probably set to a non-zero level (as compared to "no signal"). A pre-amplifier will typically convert the current signal from the photodiode into a voltage signal, which may be filtered to clean it up

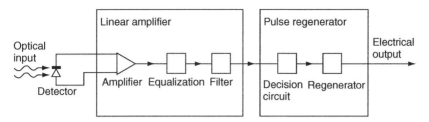

Figure 3.18   Block diagram of a digital optical communication receiver.

a bit. A pulse regenerator with a decision circuit will convert this into a digital signal that can be used by electronics (Figure 3.18).

There are three major sources of regeneration error. The first is simply noise. The receiver noise level increases with the bandwidth.

The second source of regeneration error is timing errors. The decision circuit samples the signal, and any timing variations which affect the moment of arrival of the signal, or the moment of the sampling within the decision circuit may cause the waveform to be sampled at a point other than at its maximum amplitude. The proper point in the pulse cycle to sample the signal is determined by a **clock recovery circuitry** using phase locked loops. The narrower the pulse, the more serious such variations are.

The third source of regeneration error is intersymbol interference, which occurs if the power received during one-bit period affects the signal amplitude during any other bit period. The lower the amplifier bandwidth, the more likely this is, because the pulse response is more spread out, and extends into adjacent bit periods. Electronic filters can be designed to minimize this problem.

The amplification stages of the receiver amplify both the signal and the noise. There are various ways to minimize the noise by narrowing the receiver bandwidth or performing signal averaging, including boxcar integration, multichannel scaling, pulse height analysis, and phase-sensitive detection. For example, boxcar averaging takes its name from gating the signal detection time into a repetitive train of N pulse intervals, or "boxes," during which the signal is present. Since noise which would have been accumulated during times when the gating is off is eliminated, this process improves the SNR by a factor of the square root of N for white noise, Johnson noise, or shot noise. (This is because the integrated signal contribution increases as N, while the noise contribution increases only as the square root of N.) [6] Narrowing the bandwidth of a fiber optic receiver can also have beneficial effects in controlling RIN noise (see Section 4.3.3) and modal noise (see Section 4.3.5). The use of differential signaling is also common in datacom receiver circuits, although they typically do not use lock-in amps but rather solid state electronics such as operational amplifiers.

3.5.1   SHOT NOISE

Shot noise occurs in all types of radiation detectors. It is due to the quantum nature of photoelectrons. Since individual photoelectrons are created by absorbed photons at random intervals, the resulting signal has some variation with time. The variation of detector current with time appears as noise; this can be due to either the desired signal photons or by background flux (in the latter case, the detector is said to operate in a **Background Limited in Performance**, or BLIP, mode). To study the shot noise in a photodiode, we will consider the photodetection process. An optical signal and background radiation are absorbed by the photodiode, whereby electron-hole pairs are generated. These electrons and holes are then separated by the electric field and drift toward the opposite sides of the p-n junction. In the process, a displacement current is induced in the external load resister. The photocurrent generated by the optical signal is I_p. The current generated by the background radiation is I_b. The current generated by the thermal generation of electron-hole pairs under completely dark environment is I_d. Because of the randomness of the generation of all these currents, they contribute shot noise given by a mean square current variation of

$$I_s^2 = 2q(I_p + I_b + I_d)B \qquad (3.28)$$

where q is the charge of an electron (1.6×10^{-19} coulomb) and B is the bandwidth. The equation shows that shot noise increases with current and with bandwidth. Shot noise is at its minimum when only dark current exists, and it increases with the current resulting from optical input.

3.5.2   THERMAL NOISE

Also known as **Johnson or Nyquist noise**, thermal noise is caused by randomness in carriers generation and recombination due to thermal excitation in a conductor; it results in fluctuations in the detector's internal resistance, or in any resistance in series with the detector. These resistances consist of R_j the junction resistance, R_s the series resistance, R_l the load resistance, and R_i the input resistance of the following amplifier. All the resistances contribute additional thermal noise to the system.

The series resistance R_s is usually much smaller than the other resistance and can be neglected. The thermal noise is given by

$$I_t^2 = 4kTB \left(\frac{1}{R_j} + \frac{1}{R_l} + \frac{1}{R_i} \right) \tag{3.29}$$

where k is Boltzmann's constant (1.38×10^{-23} J/K), T is absolute temperature (kelvin scale), and B is the bandwidth.

3.5.3 OTHER NOISE SOURCES

Generation-recombination noise and **1/f noise** are particular to photoconductors. Absorbed photons can produce both positive and negative charge carriers, some of which may recombine before being collected. Generation recombination noise is due to the randomness in the creation and cancellation of individual charge carriers. It can be shown [4] that the magnitude of this noise is given by

$$I_{gr} = 2I\sqrt{(\tau B/N(1 + (2\pi f\tau)^2))} = 2qG\sqrt{\varepsilon EAB} \tag{3.30}$$

where I is the average current due to all sources of carriers (not just photocarriers), τ is the carrier lifetime, N is the total number of free carriers, f is the frequency at which the noise is measured, G is the photoconductive gain (number of electrons generated per photogenerated electron), E is the photon irradiance, and A is the detector active area.

Flicker or the so-called "1/f noise" is particular to biased conductors. Its cause is not well understood, but it is thought to be connected to the imperfect conductive contact at detector electrodes. It can be measured to follow a curve of $1/f^\beta$, where β is a constant which varies between 0.8 and 1.2; the rapid falloff with $1/f$ gives rise to the name. Lack of good ohmic contact increases this noise, but it is not known if any particular type of electrical contact will eliminate this noise. The empirical expression for the noise current is

$$I_f = \alpha \left(\frac{iB}{f^\beta} \right)^{1/2} \tag{3.31}$$

where α is a proportionality constant, i is the current through the detector, and the exponents are empirically estimated to be $\alpha = 2$ and $\beta = 1$. Note this is only an empirical expression for a poorly understood phenomena; the noise current does not become infinite as f approaches zero (DC operation).

There may be other noise sources in the detector circuitry, as well; these can also be modeled as equivalent currents. The noise sources described here are uncorrelated and thus must be summed as RMS values rather than a linear summation (put another way, they add in quadrature or use vector addition), so that the total noise is given by

$$I_{tot}^2 = I_f^2 + I_{gr}^2 + I_s^2 + I_t^2 \tag{3.32}$$

Usually, one of the components in the above expression will be the dominant noise source in a given application. When designing a data link, one must keep in mind likely sources of noise, their expected contribution, and how to best reduce them. Choose a detector so that the signal will be significantly larger than the detector's expected noise. In a laboratory environment, cooling some detectors will minimize the dark current; this is not practical in most applications. There are other rules of thumb which can be applied to specific detectors, as well. For example, although Johnson noise cannot be eliminated, it can be minimized. In photodiodes, the shot noise is approximately 3X greater than Johnson noise if the DC voltage generated through a transimpedance amplifier is more than about 500 mV. This results in a higher degree of linearity in the measurements while minimizing thermal noise. The same rule of thumb applies to PbS and PbSe-based detectors, though care should be taken not to exceed the maximum bias voltage of these devices or catastrophic breakdown will occur. For the measured values of NEP, detectivity, and specific detectivity to be meaningful, the detector should be operating in a high impedance mode so that the principle source of noise is the shot noise associated with the dark current and the signal current. Although it is possible to use electronic circuits to filter out some types of noise, it is better to have the signal much stronger than the noise by either having a strong signal level or a low noise level. Several types of noise are associated with the photodiode itself and with the receiver; for example, we have already mentioned multiplication noise in an APD, which arises because multiplication varies around a statistical mean.

3.5.4 SIGNAL-TO-NOISE RATIO

Signal-to-noise ratio is a way to describe the quality of signals in a communication system. SNR is simply the ratio of the average signal power, S, to the average noise power, N, from all noise sources.

$$\text{SNR} = \frac{S}{N} \tag{3.33}$$

SNR can also be written in decibels as

$$\text{SNR} = 10\log_{10}\left(\frac{S}{N}\right) \tag{3.34}$$

If the signal power is 20 mW and the noise power is 20 nW, the SNR ratio is 1000, or 30 dB. A large SNR means that the signal is much larger than the noise. The signal power depends on the power of the incoming optical power. The specification for SNR is dependent on the application requirements.

For digital systems, **bit error rate (BER)** usually replaces SNR as a performance indicator of system quality. BER is the ratio of incorrectly transmitted bits to correctly transmitted bits. A ratio of 10^{-10} means that one wrong bit is received for every 10 billion bits transmitted. Similarly with SNR, the specification for BER is also dependent on the application requirements. SNR and BER are related. In a ideal system, a better SNR should also have a better BER. However, BER also depends on data-encoding formats and receiver designs. There are techniques to detect and correct bit errors. We can not easily calculate the BER from the SNR, because the relationship depends on several factors, including circuit design and bit error correction techniques. For a more complete overview of BER and other sources of error in a fiber optic data link, see Chapter 7.

References

[1] Sher DeCusatis, C.J. (1997) "Fundamentals of Detectors", in *Handbook of Applied Photometry*", C. DeCusatis (ed.), Chapter 3, pp. 101–132, AIP Press: N.Y.

[2] Burns, G. (1985) *Solid State Physics*, Chapter 10, pp. 323–324, Orlando, Fl.:Academic Press.

[3] NBS Special Publication SP250-17, "The NBS Photodetector Spectral Response Calibration Transfer Program".

[4] Lee, T.P. and Li, T. (1979) "Photodetectors", in *Optical Fiber Communications*, S.E. Miller and A.G. Chynoweth (eds) Chapter 18, Academic Press: N.Y.

[5] Gowar, J. (1984) *Optical Communication Systems*, Prentice Hall, Englewood Cliffs, N.J.

[6] Dereniak, E. and Crowe, D. (1984) *Optical Radiation Detectors*, John Wiley & Sons: N.Y.

[7] Graeme, J. (1994) "Divide and conquer noise in photodiode amplifiers", *Electronic Design*, pp. 10–26, 27 June.

[8] Graeme, J. (1994) "Filtering cuts noise in photodiode amplifiers", *Electronic Design*, pp. 9–22, 7 November.

[9] Graeme, J. (1987) "FET op amps convert photodiode outputs to usable signals", EDN, p. 205, 29 October.

[10] Graeme, J. (1982) "Phase compensation optimizes photodiode bandwidth", EDN, p. 177, 7 May.

[11] Bell, D.A. (1985) *Noise and the Solid State*, John Wiley & Sons: N.Y.

[12] Burt, R. and Stitt, R. (1988) "Circuit lowers photodiode amplifier noise", EDN, p. 203, 1 September.

[13] Garmire, E. (2001) "Sources, Modulators and Detectors for Fiber Optic Communication Systems", *Handbook of Optics IV*, McGraw-Hill.

[14] Cohen, Marshall J. (2000) "Photodiode arrays help meet demand for WDM", *Optoelectronics World*, Supplement to Laser Focus World, August.

[15] M. Selim Ünlü and Samuel Strite (1995) "Resonant Cavity Enhanced (RCE) Photonic Devices", at http://photon.bu.edu/selim/papers/apr-95/node1.html.

Chapter 4 | Fiber Optic Link Design

In this chapter, we will examine the technical requirements for designing fiber optic communication systems. We begin by defining some figures of merit to characterize the system performance. Then, concentrating on digital optical communication systems, we will describe how to design an optical link loss budget and how to account for various types of noise sources in the link.

4.1 Figures of Merit

There are several possible figures of merit which may be used to characterize the performance of an optical communication system. Different figures of merit may be more suitable for different applications, such as analog or digital transmission. Even if we ignore the practical considerations of laser eye safety standards, an optical transmitter is capable of launching a limited amount of optical power into a fiber; similarly, there is a limit as to how weak a signal can be detected by the receiver in the presence of noise. Thus, a fundamental consideration in optical communication systems design is the optical link **power budget**, or the difference between the transmitted and the received optical power levels. Some power will be lost due to connections, splices, and attenuation in the fiber; there may also be optical power penalties due to dispersion or other effects which we will describe later. The optical power levels define the **SNR** at the receiver, which is often used to characterize the performance of analog communication systems (such as some types of cable television). For digital transmission, the most common figure of merit is the **BER**, which is the ratio of received bit errors to the total number of transmitted bits. SNR is related to the BER by the equation

$$BER = \frac{1}{\sqrt{2\pi}} \int_Q^\infty e^{-\frac{Q^2}{2}} dQ \cong; \frac{1}{Q\sqrt{2\pi}} e^{-\frac{Q^2}{2}} \tag{4.1}$$

where Q represents the SNR for simplicity of notation [1–4]. A plot of BER vs. received optical power yields a straight line on a semilog scale, as illustrated in Figure 4.1. Some effects, such as fiber attenuation, can be overcome by increasing the received optical power, subject to constraints on maximum optical power (laser eye safety) and the limits of receiver sensitivity. Other types of noise sources are independent of signal strength. When such noise is present, no amount of increase in transmitted signal strength will affect the BER; a noise floor is produced, as shown by curve B in Figure 4.1. This type of noise can be a serious limitation on link performance. If we plot BER vs. receiver sensitivity for increasing optical power, we obtain a curve similar to Figure 4.2 which shows that for very high power levels, the receiver will go into

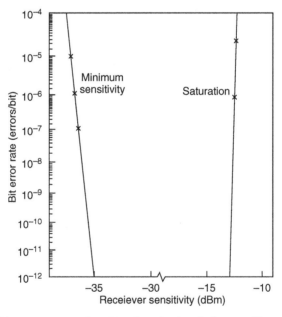

Figure 4.1   Bit error rate as a function of received optical power illustrating range of operation from minimum sensitivity to saturation [3].

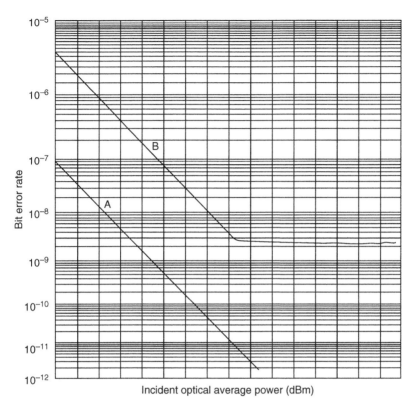

Figure 4.2 Bit error rate as a function of received optical power. Curve A shows typical performance, whereas curve B shows a BER floor [3].

saturation. The characteristic "bathtub" shaped curve illustrates a window of operation with both upper and lower limits on the received power.

In the design of some analog optical communication systems, as well as some digital television systems (e.g. those based on 64 bit Quadrature Amplitude Modulation), another possible figure of merit is the **modulation error ratio (MER)**. To understand this metric, consider the standard definition given by the Digital Video Broadcasting (DVB) Measurements Group [5]. First, the video receiver captures a time record of N received signal coordinate pairs, representing the position of information on a two-dimensional screen. The ideal position coordinates are given by the vector (X_j, Y_j). For each received symbol, a decision is made as to which symbol was transmitted, and an error vector $(\Delta X_j, \Delta Y_j)$ is defined as the distance from the ideal position to the actual position of the received symbol. The MER is then defined as the sum of the squares of the magnitudes of the ideal symbol vector divided by the sum of the squares of the magnitudes of the symbol error vectors:

$$MER = 10 \log \frac{\sum_{j=1}^{N}(X_j^2 + Y_j^2)}{\sum_{j=1}^{N}(\Delta X_j^2 + \Delta Y_j^2)} \quad \text{dB.} \tag{4.2}$$

When the signal vectors are corrupted by noise, they can be treated as random variables. The denominator becomes an estimate of the average power of the error vector and contains all signal degradation due to noise, reflections, etc. If the only significant source of signal degradation is additive white Gaussian noise, then MER and SNR are equivalent. For communication systems which contain other noise sources, MER offers some advantages; in particular, for some digital transmission systems there may be a very sharp change in BER as a function of SNR (the so-called "cliff effect") which means that BER alone cannot be used as an early predictor of system failures. MER, on the other hand, can be used to measure signal-to-interference ratios accurately for such systems. Because MER is a statistical measurement, its accuracy is directly related to the number of vectors, N, used in the computation; good accuracy can be obtained with $N = 10,000$, which would require about 2 ms to accumulate at the industry standard digital video rate of 5.057 Msymbols/s.

Analog fiber optic links are in general nonlinear. That is, if the input electrical information is a harmonic signal of frequency f_0, the output electrical signal will contain the fundamental frequency f_0 as well as high-order harmonics of frequencies nf_0 ($n > 2$). These high-order harmonics comprise the harmonic distortions of analog fiber optic links [6]. The nonlinear behavior is caused by nonlinearities in the transmitter, the fiber, and the receiver.

The same sources of nonlinearities in the fiber optic links lead to **Inter-Modulation Distortions (IMD)**, which can be best illustrated in a two-tone transmission scenario. If the input electrical information is a superposition of two harmonic signals of frequencies f_1 and f_2, the output electrical signal will contain second-order intermodulation at frequencies $f_1 + f_2$ and $f_1 - f_2$ as well as third-order intermodulation at frequencies $2f_1 - f_2$ and $2f_2 - f_1$.

Most analog fiber optic links require bandwidth of less than one octave ($f_{\max} < 2f_{\min}$). As a result harmonic distortions as well as second-order IMD products are not important as they can be filtered out electronically. However, third-order IMD products are in the same frequency range (between $f_{\min}$ and $f_{\max}$) as the signal itself and therefore appear in the output signal as the spurious response. Thus the linearity of analog fiber optic links is determined by the level of third-order IMD products. In the case of analog links where third-order IMD is eliminated through linearization circuitry, the lowest odd-order IMD determines the linearity of the link.

To quantify IMD distortions, a two-tone experiment (or simulation) is usually conducted where the input radio frequency (RF) powers of the two tones are equal. There are several important features to bear in mind when interpreting the linear and nonlinear power transfer functions (the output RF power of each of two input tones and the second or third-order IMD product as a function of the input RF power of each input harmonic signal). When plotted on a log-log scale, the fundamental power transfer function should be a line with a slope of unity. The second- (third-) order power transfer function should be a line with a slope of two (three). The intersections of the power transfer functions are called second- and third-order intercept points, respectively. Because of the fixed slopes of the power transfer functions, the intercept points can be calculated from measurements obtained at a single input power level. At a certain input level, the output power of each of the two fundamental tones, the second-order IMD product, and third-order IMD products are P_1, P_2, and P_3, respectively. When the power levels are in units of dB or dBm, the second-order and third-order intercept points are

$$IP_2 = 2P_1 - P_2 \qquad (4.3)$$

and

$$IP_3 = \frac{3P_1 - P_3}{2}. \qquad (4.4)$$

The dynamic range is a measure of the ability of an analog fiber optic link to faithfully transmit signals at various power levels. At the low input power end, the analog link can fail due to insufficient power level so that the output power is below the noise level. At the high input power end, the analog link can fail due to the fact that the IMD products become the dominant source of signal degradation. In terms of the output power, the dynamic range (of the output power) is defined as the ratio of the fundamental output to the noise power. However, it should be noted that the third-order IMD products increase three times faster than the fundamental signal. After the third-order IMD products exceeds the noise floor, the ratio of the fundamental output to the noise power is meaningless as the dominant degradation of the output signal comes from IMD products. So a more meaningful definition of the dynamic range is the so-called **spurious-free dynamic range (SFDR)** [6, 7], which is the ratio of the fundamental output to the noise power at the point where the IMD products is at the noise level. The SFDR is then practically the maximum dynamic range. Since the noise floor depends on the bandwidth of interest, the unit for SFDR should be $dBHz^{2/3}$. The dynamic range decreases as the bandwidth of the system is increased. The SFDR is also often defined with reference to the input power, which corresponds to SFDR with reference to the output power if there is no gain compression.

4.2 Link Budget Analysis

In order to design a proper optical data link, the contribution of different types of noise sources should be assessed when developing a link budget. There are two basic approaches to link budget modeling. One method is to design the link to operate at the desired BER when all the individual link components assume their worst case performance. This conservative approach is desirable when very high performance is required, or when it is difficult or inconvenient to replace failing components near the end of their useful lifetimes. The resulting design has a high safety margin; in some cases, it may be overdesigned for the required level of performance. Since it is very unlikely that all the elements of the link will assume their worst case performance at the same time, an alternative is to model the link budget statistically. For this method, distributions of transmitter power output, receiver sensitivity, and other parameters are either measured or estimated. They are then combined statistically using an approach such as the Monte Carlo method, in which many possible link combinations are simulated to generate an overall distribution of the available link optical power. A typical approach is the **3-sigma design**, in which the combined variations of all link components are not allowed to extend more than 3 standard deviations from the average performance target in either direction. The statistical approach results in greater design flexibility, and generally increased distance compared with a worst-case model at the same BER.

4.2.1 INSTALLATION LOSS

It is convenient to break down the link budget into two areas: installation loss and available power. **Installation loss** refers to optical losses associated with the fiber cable, such as connector loss and splice loss.

Available optical power is the difference between the transmitter output and the receiver input powers, minus additional losses due to noise sources. With this approach, the installation loss budget may be treated statistically and the available power budget as worst case. First, we consider the installation loss budget, which can be broken down into three areas, namely transmission loss, fiber attenuation as a function of wavelength, and connector or splice losses.

4.2.2   TRANSMISSION LOSS

Transmission loss is perhaps the most important property of an optical fiber; it affects the link budget and maximum unrepeated distance. Since the maximum optical power launched into an optical fiber is determined by international laser eye safety standards [8], the number and separation between optical repeaters and regenerators is largely determined by this loss. The mechanisms responsible for this loss include material absorption as well as both linear and nonlinear scattering of light from impurities in the fiber [1–5]. Typical loss for single-mode optical fiber is about 2–3 dB/km near 800 nm wavelength, 0.5 dB/km near 1300 nm, and 0.25 dB/km near 1550 nm. Multimode fiber loss is slightly higher, and bending loss will only increase the link attenuation further.

4.2.3   ATTENUATION VS. WAVELENGTH

Since fiber loss varies with wavelength, changes in the source wavelength or use of sources with a spectrum of wavelengths will produce additional loss. Transmission loss is minimized near the 1550 nm wavelength band, which unfortunately does not correspond with the dispersion minimum at around 1310 nm. An accurate model for fiber loss as a function of wavelength is quite complex, and must include effects such as linear scattering, macrobending, material absorption due to ultraviolet and infrared band edges, hydroxide (OH) absorption, and absorption from common impurities such as phosphorous. Typical loss due to wavelength dependent attenuation for laser sources on single-mode fiber is specified by the manufacturer and can be held below 0.1 dB/km.

4.2.4   CONNECTOR AND SPLICE LOSS

There are also installation losses associated with fiber optic connectors and splices; both of these are inherently statistical in nature. There are many different kinds of standardized optical connectors, and different models which have been published for estimating connection loss due to fiber misalignment [9, 10]; most of these treat loss due to misalignment of fiber cores, offset of fibers on either side of the connector, and angular misalignment of fibers. There is no general model available to treat all types of connectors, but typical connector loss values average about 0.5 dB worst case for multimode, slightly higher for single mode.

Optical splices are required for longer links, since fiber is usually available in spools of 1–5 km, or to repair broken fibers. There are two basic types: mechanical splices (which involve placing the two fiber ends in a receptacle that holds them close together, usually with epoxy) and the more commonly used fusion splices (in which the fiber are aligned, then heated sufficiently to fuse the two ends together). Typical splice loss values can be kept well below 0.2 dB.

4.3   Optical Power Penalties

Next, we will consider the assembly loss budget, which is the difference between the transmitter output and the receiver input powers, allowing for optical power penalties due to noise sources in the link. We will follow the standard convention of assuming a digital optical communication link which is best characterized by its BER. Contributing factors to link performance include the following:

- Dispersion (modal and chromatic) or intersymbol interference
- Mode partition noise
- Multipath interference
- Relative intensity noise
- Timing jitter
- Radiation induced darkening
- Modal noise.

Higher order, nonlinear effects including Stimulated Raman and Brillouin scattering and frequency chirping are important for long haul links or optical amplifiers, but are quite complex and beyond the scope of this book.

4.3.1   DISPERSION

The most important fiber characteristic after transmission loss is dispersion, or intersymbol interference. This refers to the broadening of optical pulses as they propagate along the fiber. As pulses broaden, they tend to interfere with adjacent pulses; this limits the maximum achievable data rate. In multimode fibers, there are two dominant kinds of dispersion, modal and chromatic. **Modal dispersion** refers to the fact that different modes will travel at different velocities and cause pulse broadening. The fiber's **modal bandwidth** in units of MHz-km is specified according to the expression

$$BW_{modal} = BW_1/L^{\gamma}, \tag{4.5}$$

where BW_{modal} is the modal bandwidth for a length L of fiber, BW_1 is the manufacturer-specified modal bandwidth of a 1 km section of fiber, and γ is a constant known as the modal bandwidth concatenation length scaling factor. The term γ usually assumes a value between 0.5 and 1, depending on details of the fiber manufacturing and design as well as the operating wavelength; it is conservative to take $\gamma = 1.0$.

The other major contribution is chromatic dispersion, BW_{chrom}, which occurs because different wavelengths of light propagate at different velocities in the fiber. For multimode fiber, this is given by

$$BW_{chrom} = \frac{L^{\gamma_c}}{(\sqrt{\lambda_w}(a_0 + a_1|\lambda_c - \lambda_{eff}|)}, \tag{4.6}$$

where L is the fiber length in km; λ_c is the center wavelength of the source in nm; λ_w is the source FWHM spectral width in nm; γ_c is the chromatic bandwidth length scaling coefficient, a constant; λ_{eff} is the effective wavelength, which combines the effects of the fiber zero dispersion wavelength and spectral loss signature; and the constants a_1 and a_0 are determined by a regression fit of measured data. The chromatic bandwidth for 62.5/125 micron fiber is empirically given by [11]

$$BW_{chrom} = \frac{10^4 L^{-0.69}}{\sqrt{\lambda_w}(1.1 + 0.0189|\lambda_c - 1370|)}. \tag{4.7}$$

For this expression, the center wavelength was 1335 nm and λ_{eff} was chosen midway between λ_c and the water absorption peak at 1390 nm; although λ_{eff} was estimated in this case, the expression still provides a good fit to the data. For 50/125 micron fiber, the expression becomes

$$BW_{chrom} = \frac{10^4 L^{-0.65}}{\sqrt{\lambda_w}(1.01 + 0.0177|\lambda_c - 1330|)}. \tag{4.8}$$

For this case, λ_c was 1313 nm and the chromatic bandwidth peaked at $\lambda_{eff} = 1330$ nm. Recall that this is only one possible model for fiber bandwidth [1]. The total bandwidth capacity of multimode fiber BW_t is obtained by combining the modal and chromatic dispersion contributions, according to

$$\frac{1}{BW_t^2} = \frac{1}{BW_{chrom}^2} + \frac{1}{BW_{modal}^2}. \tag{4.9}$$

Once the total bandwidth is known, the dispersion penalty can be calculated for a given data rate. One expression for the dispersion penalty in dB is

$$P_d = 1.22 \left[\frac{Bit\ Rate\ (Mb/s)}{BW_t\ (MHz)} \right]^2. \tag{4.10}$$

For typical telecommunication grade fiber, the dispersion penalty for a 20 km link is about 0.5 dB.

Dispersion is usually minimized at wavelengths near 1310 nm; special types of fiber have been developed which manipulate the index profile across the core to achieve minimal dispersion near 1550 nm, which is also the wavelength region of minimal transmission loss. Unfortunately, this dispersion-shifted fiber suffers from some practical drawbacks, including susceptibility to certain kinds of nonlinear noise and increased interference between adjacent channels in a wavelength multiplexing environment. There is a new type of fiber which minimizes dispersion while reducing the unwanted crosstalk effects, called **dispersion optimized fiber**. By using a very sophisticated fiber profile, it is possible to minimize dispersion over the entire wavelength range from 1300 to 1550 nm, at the expense of very high loss (around 2 dB/km); this is known as **dispersion flattened fiber**. Yet another approach is called **dispersion compensating fiber**; this fiber is designed with negative dispersion characteristics, so that when used in series with conventional fiber it will "undisperse" the signal. Dispersion compensating fiber has a much narrower core than standard single-mode fiber, which makes it susceptible to nonlinear effects; it is also birefringent and suffers from polarization mode dispersion, in which different states of polarized light propagate with very different group velocities.

By definition, single-mode fiber does not suffer modal dispersion. **Chromatic dispersion** is an important effect, though, even given the relatively narrow spectral width of most laser diodes. The dispersion of single-mode fiber is given by

$$D = \frac{d\tau_g}{d\lambda} = \frac{S_o}{4} \left(\lambda_c - \frac{\lambda_o^4}{\lambda_c^3} \right), \tag{4.11}$$

where D is the dispersion in ps/(km-nm) and λ_c is the laser center wavelength. The fiber is characterized by its zero dispersion wavelength, λ_o, and zero dispersion slope, S_o. Usually, both center wavelength and zero dispersion wavelength are specified over a range of values; it is necessary to consider both upper and lower bounds in order to determine the worst case dispersion penalty. Once the dispersion is determined, the intersymbol interference penalty as a function of link length, L, can be determined to a good approximation from [12]:

$$P_d = 5 \log \left(1 + 2\pi \left(BD\Delta\lambda\right)^2 L^2\right), \qquad (4.12)$$

where B is the bit rate and $\Delta\lambda$ is the RMS spectral width of the source. By maintaining a close match between the operating and the zero dispersion wavelengths, this penalty can be kept to a tolerable 0.5–1.0 dB in most cases.

4.3.2 MODE PARTITION NOISE

This penalty is related to the properties of a FP type laser diode cavity; although the total optical power output from the laser may remain constant, the optical power distribution among the laser's longitudinal modes will fluctuate. We must be careful to distinguish this behavior of the instantaneous laser spectrum, which varies with time, from the time-averaged spectrum which is normally observed experimentally. The light propagates through a fiber with wavelength dependent dispersion or attenuation, which deforms the pulse shape. Each mode is delayed by a different amount due to group velocity dispersion in the fiber; this leads to additional signal degradation at the receiver, known as mode partition noise; it is capable of generating BER floors, such that additional optical power into the receiver will not improve the link BER. The power penalty due to mode partition noise can be estimated from [13]:

$$P_{mp} = 5 \log(1 - Q^2 \sigma_{mp}^2), \qquad (4.13)$$

where

$$\sigma_{mp}^2 = \frac{1}{2} k^2 (\pi B)^4 [A_1^4 \Delta\lambda^4 + 42 A_1^2 A_2^2 \Delta\lambda^6 + 48 A_2^4 \Delta\lambda^8], \qquad (4.14)$$

$$A_1 = DL, \qquad (4.15)$$

and

$$A_2 = \frac{A_1}{2(\lambda_c - \lambda_o)}. \qquad (4.16)$$

The mode partition coefficient, k, is a number between 0 and 1 which describes how much of the optical power is randomly shared between modes; it summarizes the statistical nature of mode partition noise. While there are significantly more complex models available and work in this area is ongoing, a practical rule of thumb is to keep the mode partition noise penalty less than 1.0 dB maximum, provided that this penalty is far away from any noise floors.

4.3.3 RELATIVE INTENSITY NOISE

Stray light reflected back into a FP type laser diode gives rise to intensity fluctuations in the laser output. This is a complicated phenomena, strongly dependent on the type of laser; it is called either reflection-induced intensity noise or RIN. This effect is important since it can also generate BER floors. The power penalty due to RIN is the subject of ongoing research; since the reflected light is measured at a specified signal level, RIN is data dependent although it is independent of link length. Since many laser diodes are packaged in windowed containers, it is difficult to correlate the RIN measurements on an unpackaged laser with those of a commercial product. There have been several detailed attempts to characterize RIN [14, 15]; typically, the RIN noise is assumed Gaussian in amplitude and uniform in frequency over the receiver bandwidth of interest. The RIN value is specified for a given laser by measuring changes in the optical power when a controlled amount of light is fed back into the laser; it is signal dependent, and is also influenced by temperature, bias voltage, laser structure, and other factors which typically influence laser output power [15]. If we assume that the effect of RIN is to produce an equivalent noise current at the receiver, then the additional receiver noise σ_r may be modeled as

$$\sigma_r = \gamma^2 S^{2g} B, \tag{4.17}$$

where S is the signal level during a bit period, B is the bit rate, and g is a noise exponent which defines the amount of signal dependent noise. If $g = 0$, noise power is independent of the signal, while for $g = 1$ noise power is proportional to the square of the signal strength. The coefficient γ is given by

$$\gamma^2 = S_i^{2(1-g)} 10^{(RIN_i/10)}, \tag{4.18}$$

where RIN_i is the measured RIN value at the average signal level S_i, including worst case backreflection conditions and operating temperatures. One approximation for the RIN power penalty is given by

$$P_{rin} = -5 \log \left[1 - Q^2(BW)(1 + M_r)^{2g} \left(10^{RIN/10} \right) \left(\frac{1}{M_r} \right)^2 \right], \tag{4.19}$$

where the RIN value is specified in dB/Hz, BW is the receiver bandwidth, M_r is the receiver modulation index, and the exponent g is a constant varying between 0 and 1 which relates the magnitude of RIN noise to the optical power level. The maximum RIN noise penalty in a link can usually be kept to below 0.5 dB.

4.3.4 JITTER

Although it is not strictly an optical phenomena, another important area in link design deals with the effects of timing jitter on the optical signal. In a typical optical link, a clock is extracted from the incoming data signal which is used to retime and reshape the received digital pulse; the received pulse is then compared with a threshold to determine if a digital "1" or "0" was transmitted. So far, we have discussed BER testing with the implicit assumption that the measurement was made in the center of the received data bit; to achieve this, a clock transition at the center of the bit is required. When the clock is generated from a receiver timing recovery circuit, it will have some variation in time and the exact location of the clock edge will be uncertain. Even if the clock is positioned at the center of the bit, its position may drift over time. There will be a region of the bit interval, or eye, in the time domain where the BER is acceptable; this region is defined as the eyewidth [1–3]. Eyewidth measurements are an important parameter for evaluation of fiber optic links; they are intimately related to the BER, as well as the acceptable clock drift, pulse width distortion, and optical power. At low optical power levels, the receiver SNR is reduced; increased noise causes amplitude variations in the received signal. These amplitude variations are translated into time domain variations in the receiver decision circuitry, which narrows the eyewidth. At the other extreme, an optical receiver may become saturated at high optical power, reducing the eyewidth and making the system more sensitive to timing jitter. This behavior results in the typical "bathtub" curve shown in Figure 4.2; for this measurement, the clock is delayed from one end of the bit cell to the other, with the BER calculated at each position. Near the ends of the cell, a large number of errors occur; towards the center of the cell, the BER decreases to its true value. The eye opening may be defined as the portion of the eye for which the BER remains constant; pulse width distortion occurs near the edges of the eye, which denotes the limits of the valid clock timing. Uncertainty in the data pulse arrival times causes errors to occur by closing the eye window and causing the eye pattern to be sampled away from the center. This is one of the fundamental problems of optical and digital signal processing, and a large body of work has been done in this area [16, 17]. In general, multiple jitter sources will be present in a link; these will tend to be uncorrelated.

Industry standards bodies have adopted a definition of jitter [17] as short-term variations of the significant instants (rising or falling edges) of a digital signal from their ideal position in time. Longer-term variations are described as wander; in terms of frequency, the distinction between jitter and wander is somewhat unclear. Each component of the optical link (data source, serializer, transmitter, encoder, fiber, receiver, retiming/clock recovery/deserialization, decision circuit) will contribute some fraction of the total system jitter. If we consider the link to be a "black box" (but not necessarily a linear system) then we can measure the level of output jitter in the absence of input jitter; this is known as the "intrinsic jitter" of the link. The relative importance of jitter from different sources may be evaluated by measuring the spectral density of the jitter. Another approach is the maximum tolerable input jitter (MTIJ) for the link. Finally, since jitter is essentially a stochastic process, we may attempt to characterize the jitter transfer function (JTF) of the link, or estimate the probability density function of the jitter. The problem of jitter accumulation in a chain of repeaters becomes increasingly complex; however, we can state some general rules of thumb. For well designed practical networks, the basic results of jitter modeling remain valid for N nominally identical repeaters transmitting random data; systematic jitter accumulates in proportion to $N^{1/2}$ and random jitter accumulates in proportion to $N^{1/4}$. For most applications, the maximum timing jitter should be kept below about 30% of the maximum receiver eye opening.

4.3.5 MODAL NOISE

Because high capacity optical links tend to use highly coherent laser transmitters, random coupling between fiber modes causes fluctuations in the optical power coupled through splices and connectors; this phenomena is known as modal noise [18]. As one might expect, modal noise is worst when using laser sources in conjunction with multimode fiber; recent industry standards have allowed the use of short-wave lasers (750–850 nm) on 50 micron fiber which may experience this problem. Modal noise is usually considered to be nonexistent in single-mode systems. However, modal noise in single-mode fibers can arise when higher order modes are generated at imperfect connections or splices. If the lossy mode is not completely attenuated before it reaches the next connection, interference with the dominant mode may occur. For N sections of fiber, each of length L in a single-mode link, the worst case sigma for modal noise can be given by

$$\sigma_m = \sqrt{2}\, N\eta(1 - \eta)e^{-\alpha L}, \tag{4.20}$$

where α is the attenuation coefficient of the LP_{11} mode, and η is the splice transmission efficiency, given by

$$\eta = 10^{-(\eta_o/10)}, \tag{4.21}$$

where η_o is the mean splice loss (typically, splice transmission efficiency will exceed 90%). The corresponding optical power penalty due to modal noise is given by

$$P = -5\log(1 - Q^2\sigma_m^2), \tag{4.22}$$

where Q corresponds to the desired BER. This power penalty should be kept to less than 0.5 dB.

4.3.6 RADIATION INDUCED LOSS

Another important environmental factor as mentioned earlier is exposure of the fiber to ionizing radiation damage. There is a large body of literature concerning the effects of ionizing radiation on fiber links [19, 20]. There are many factors which can affect the radiation susceptibility of optical fiber, including the type of fiber, type of radiation (gamma radiation is usually assumed to be representative), total dose, dose rate (important only for higher exposure levels), prior irradiation history of the fiber, temperature, wavelength, and data rate. Almost all commercial fiber is intentionally doped to control the refractive index of the core and cladding, as well as dispersion properties. Because of the many factors involved, there does not exist a comprehensive theory to model radiation damage in optical fibers, although the basic physics of the interaction has been described. There are two dominant mechanisms: radiation induced darkening and scintillation. First, high energy radiation can interact with dopants, impurities, or defects in the glass structure to produce color centers which absorb strongly at the operating wavelength. Carriers can also be freed by radiolytic or photochemical processes; some of these become trapped at defect sites, which modifies the band structure of the fiber and causes strong absorption at infrared wavelengths. This radiation induced darkening increases the fiber attenuation; in some cases, it is partially reversible when the radiation is removed, although high levels or prolonged exposure will permanently damage the fiber. The second effect, scintillation, is caused if the radiation interacts with impurities to produce stray light. This light is generally broadband, but will tend to degrade the BER at the receiver; scintillation is a weaker effect than radiation-induced darkening. These effects will degrade the BER of a link; they can be prevented by shielding the fiber, or partially overcome by a third mechanism, called photobleaching. The presence of intense light at the proper wavelength can partially reverse the effects of darkening in a fiber. It is also possible to treat silica core fibers by briefly exposing them to controlled levels of radiation at controlled temperatures; this increases the fiber loss, but makes the fiber less susceptible to future irradiation. These so-called "radiation hardened fibers" are often used in environments where radiation is anticipated to play an important role. The loss due to normal background radiation exposure over a typical link lifetime can be held below about 0.5 dB.

References

[1] Miller, S.E. and Chynoweth, A.G. (eds) (1979) *Optical Fiber Telecommunications*, Academic Press, Inc.: New York.
[2] Gowar, J. (1984) *Optical Communication Systems*, Prentice Hall, Englewood Cliffs: N.J.
[3] DeCusatis, C., Maass, E., Clement, D. and Lasky, R. (eds) (1998) *Handbook of Fiber Optic Data Communication*, Academic Press, NY; see also *Optical Engineering*, special issue on optical data communication (December 1998).
[4] Lasky, R., Osterberg, U. and Stigliani, D. (eds) (1995) *Optoelectronics for Data Communication*, Academic Press, NY.
[5] Digital video broadcasting (DVB), Measurement Guidelines for DVB systems, European Telecommunications Standards Institute ETSI Technical Report ETR 290, May 1997; Digital Multi-Programme Systems for Television Sound and Data Services for Cable Distribution, International Telecommunications Union ITU-T Recommendation J.83, 1995; Digital Broadcasting System for Television, Sound and Data Services; Framing Structure, Channel Coding and Modulation for Cable Systems, European Telecommunications Standards Institute ETSI 300 429, 1994.
[6] Stephens, W.E. and Joseph, T.R. (1987) "System Characteristics of Direct Modulated and Externally Modulated RF Fiber-Optic Links", *IEEE J. Lightwave Technol.*, vol. LT-5 (3), pp. 380–387.
[7] Cox III, C.H. and Ackerman, E.I. (1999) "Some limits on the performance of an analog optical link", Proceedings of the SPIE–The International Society for Optical Engineering, vol. 3463, pp. 2–7.

[8] United States laser safety standards are regulated by the Dept. of Health and Human Services (DHHS), Occupational Safety and Health Administration (OSHA), Food and Drug Administration (FDA), Code of Radiological Health (CDRH), 21 Code of Federal Regulations (CFR) subchapter J; the relevant standards are ANSI Z136.1, "Standard for the safe use of lasers" (1993 revision) and ANSI Z136.2, "Standard for the safe use of optical fiber communication systems utilizing laser diodes and LED sources" (1996–97 revision); elsewhere in the world, the relevant standard is International Electrotechnical Commission (IEC/CEI) 825 (1993 revision).

[9] Gloge, D. (1975) "Propagation effects in optical fibers", *IEEE Trans. Microw. Theor. Tech.*, vol. MTT-23, pp. 106–120.

[10] Shanker, P.M. (1988) "Effect of modal noise on single-mode fiber optic network", *Opt. Comm.* 64, pp. 347–350.

[11] Refi, J.J. (1986) "LED bandwidth of multimode fiber as a function of source bandwidth and LED spectral characteristics", *IEEE J. Lightwave Tech.*, vol. LT-14, pp. 265–272.

[12] Agrawal, G.P. *et al.* (1988) "Dispersion penalty for 1.3 micron lightwave systems with multimode semiconductor lasers", *IEEE J. Lightwave Tech.*, vol. 6, pp. 620–625.

[13] Ogawa, K. (1982) "Analysis of mode partition noise in laser transmission systems", *IEEE J. Quantum Elec.*, vol. QE-18, pp. 849–855.

[14] Radcliffe, J. (1989) "Fiber optic link performance in the presence of internal noise sources", IBM Technical Report, Glendale Labs, Endicott, NY.

[15] Xiao, L.L., Su, C.B. and Lauer, R.B. (1992) "Increase in laser RIN due to asymmetric nonlinear gain, fiber dispersion, and modulation", *IEEE Photon. Tech. Lett.*, vol. 4, pp. 774–777.

[16] Trischitta, P. and Sannuti, P. (1988) "The accumulation of pattern dependent jitter for a chain of fiber optic regenerators", *IEEE Trans. Comm.*, vol. 36, pp. 761–765.

[17] CCITT Recommendations G.824, G.823, O.171, and G.703 on timing jitter in digital systems, 1984.

[18] Marcuse, D. and Presby, H.M. (1975) "Mode coupling in an optical fiber with core distortion", *Bell Sys. Tech. J.*, vol. 1, p. 3.

[19] Frieble, E.J. *et al.* (1984) "Effect of low dose rate irradiation on doped silica core optical fibers", *App. Opt.*, vol. 23, pp. 4202–4208.

[20] Haber, J.B. *et al.* (1988) "Assessment of radiation induced loss for AT&T fiber optic transmission systems in the terrestrial environment", *IEEE J. Lightwave Tech.*, vol. 6, pp. 150–154.

Chapter 5 | Repeaters and Optical Amplifiers

In this chapter, we discuss some other components commonly used in optical communication networks. In order to extend these networks to very long distances, optical amplifiers are used to increase gain. However, since the background noise is also amplified, there are limitations to this approach. Optical repeaters may be used to retime, reshape, and regenerate an optical signal, providing even longer distances than optical amplifiers used alone. Sometimes these functions are build into switches which convert the optical signal into electronic form, then retransmit the signal at a higher optical power level. Repeater and amplifier technology is often used in combination with WDM, and will be discussed in this context.

It is desirable to support the longest distance possible without repeaters between nodes in an optical communication network. Note that the total supported distance for a WDM system may depend on the number of channels in use; adding more channels requires additional wavelength multiplexing stages, and the optical fibers can reduce the available link budget. The available distance is also a function of the network topology; WDM filters may need to be configured differently, depending on whether they form an **optical seam** (configuration which does not allow a set of wavelengths to propagate into the next stage of the network) or **optical bypass** (configuration which permits wavelengths to pass through into the rest of the network). Thus, the total distance and available link loss budget in a point-to-point network may be different from the distance in a ring network. The available distance is typically independent of data rate up to around 2 Gbit/s; at higher data rates, dispersion may limit the achievable distances. This should be kept in mind when installing a new WDM system which is planned to be upgraded to significantly higher data rates in the future. The maximum available distance and link loss may also be reduced if optional devices such as switches are included in the network for protection purposes. In some cases, it is possible to concatenate or cascade optical networks together to achieve longer total distances; for example, by daisy chaining two point-to-point networks the effective distance can be doubled (if there is a suitable location in the middle of the link to house the new equipment).

Optical repeaters or switches generally convert the optical signal into electronic form, and later re-transmit the optical signal on a different fiber. We will also discuss all-optical amplifiers, which are capable of directly amplifying whatever optical signal may be present in a fiber without converting it to electronic form. Some WDM devices may also support longer distances using optical amplifiers in either **pre-amp**, **post-amp**, or **mid-span** configurations (depending, of course, on the relative location of the amplifiers). Amplifiers capable of increasing the optical power of a signal are called **optical power amplifiers**; they can be placed directly after a signal source to boost its power. Preamplifiers are placed directly in front of the receiver, and are characterized by very low noise and the ability to work with very weak signals. Mid-span or **line amplifiers** have high amplification and moderately low noise; they can be placed anywhere along the length of the fiber span. New types of distance extension technology are also under development; for example, record-breaking terabit per second transmission in systems over more than 2000 km is possible (this was done using special nonlinear optical signals called dispersion managed solitons).

Amplifiers and repeaters are needed to overcome various effects in an optical communication network. Of course, the fiber attenuates an optical signal's amplitude or intensity as it travels, and connectors or splices in the fiber path will also lower the signal level. Signal degeneration in fiber systems can also arise from various other sources, which are beyond our scope to discuss in detail but which will be mentioned here for completeness. These include nonlinear effects such as pulse spreading due to group velocity dispersion (GVD) (which can be corrected in principle through passive dispersion compensation schemes), polarization mode dispersion (PMD), and other nonlinear effects such as Kerr effect signal distortion and jitter for data rates above 10 Gbit/s. In a WDM network, when many wavelengths are present on a single fiber, the total optical power may be high enough to cause nonlinear effects, such as changing the fiber's refractive index. A particular problem in optical amplifiers is **amplified spontaneous emission (ASE)** noise, which results from the amplifier increasing the strength of both the desired signal and the undesired random noise. Not all of these effects can be overcome simply by increasing the signal strength. If timing jitter is negligible, simple amplification and reshaping processes are usually enough to maintain signal quality over long distances by preventing the accumulation of noise and distortion. Many repeaters convert a WDM optical signal to an electrical signal, then back into an optical signal for re-transmission; however, various schemes for all-optical 2R and 3R repeaters have been suggested. A 2R regenerator consists mainly of a linear amplifier (which may be an optical amplifier) followed by a data driven nonlinear optical gate (NLOG) which modulates a low noise continuous light source. If the gate transmission vs. signal intensity characteristics yield a thresholding or limiting behavior, then the signal extinction ratio can be improved and ASE noise can be partially reduced; in addition, accumulated frequency chirp of the signal can also be compensated.

In some cases, timing jitter is also a concern, for example due to cross-phase modulation in WDM systems (optical crosstalk between different communication channels) or pulse edge distortions due to the finite response times of nonlinear analog signal processing devices (wavelength converters). In these cases, 3R regeneration may be necessary. The basic structure of an optical 3R regenerator consists of an amplifier (which may be optical), a clock recovery function to provide a jitter-free short pulse clock signal, and a data driven NLOG which modulates this clock signal. The core function of the optical regenerators is thus a nonlinear gate featuring extinction ratio enhancement and noise reduction. Such gates can be either optical in nature (as in the case of proposed all-optical repeaters) or electronic (as in the case of hybrid optical-to-electrical conversion based devices). Semiconductor based gates are much more compact than fiber based devices. It is also possible to further subdivide this class of semiconductor devices into passive devices (such as saturable absorbers) and active devices (such as semiconductor optical amplifiers that require an electrical power supply). For optical 3R regeneration, a synchronous jitter-free clock stream must be recovered from the incident signal. Many solutions to this problem have been reported, and it would be beyond the scope of this chapter to describe them all. Although 2R regenerators are attractive because of their relative simplicity, it is not clear whether they will be adequate above 10 Gbit/s data rates since this would require components with very fast rise and fall time responses.

5.1   Repeaters

In order to extend optical signals over long distances, **repeaters** or **regenerators** are often used. These devices receive a modulated optical signal (typically at a high bit rate), convert it into an electronic signal at the same bit rate, amplify the signal, and re-transmit the signal in optical form. Thus, we can see that a repeater has three major parts: an optical receiver, an electronic amplifier, and an optical transmitter. Repeaters amplify only a single optical wavelength, and may be spaced every 40–50 km or so in a standard communication network (special devices are used in long haul submarine communications to extend these distances).

Better performance is usually achieved with channels that retime the data; data retiming is a desirable property, since it improves the signal fidelity and reduces noise and jitter. There are three levels of functionality, namely **Regeneration** (also known as amplification, this insures that the outgoing signal has sufficient power to reach its next destination), **Reshaping** (removes pulse shape distortion such as those caused by dispersion), and **Retiming** (removes timing jitter to improve clock recovery at the receiver). Devices which support only the first two are known as "**2R**" **repeaters**, while devices which support all three are called "**3R**" **repeaters**.* A device which only supports amplification, such as an optical amplifier, may be called a 1R device. Generally speaking, longer distances and better data fidelity are possible using 3R repeaters; however, this class of repeaters must be configured in either software, hardware, or both to recognize the data rate on the link. Care must be taken to keep the advantages of a protocol independent design when configuring a 3R repeater, since the repeater should not affect the data stream any more than necessary.

It is worth noting that repeaters can be configured to convert from multimode to single-mode fiber, allowing the use of multiple cable types. This is possible because, unlike a splice, a repeater has an optical receiver, demodulators and decision circuits, and transmitters which send the signal on to the next node. In this way, repeaters contain elements similar to those in transceivers. By choosing an appropriate transmitter for the new fiber type, a high fidelity signal can be maintained. In fact, repeaters can convert signals on copper to fiber optics as well.

---

*Note that the terminology is different for telecom submarine cables; these often refer to a "repeater" as requiring submerged electronics, while an optical amplifier is not considered a repeater at all. Thus, telecom systems may report "unrepeated" systems reaching several hundred km, and repeated systems reaching over 1000 km with repeaters spaced every 50–100 km.

5.2 Optical Amplifiers

Optical amplifiers were originally designed as a simplified alternative to repeaters, to overcome the attenuation in an optical fiber link. Similar to an electronic amplifier, they provide a certain **gain** to an optical signal (ratio of the output power to input power, measured in dB). The gain is provided by pumping a medium, which undergoes population inversion when stimulated by an optical signal. Thus, an optical amplifier functions in much the same way as a laser. These devices are characterized by their **gain efficiency**, or the gain as a function of input power (dB/mW). Only a certain range of optical frequencies, called the **gain bandwidth**, can be amplified by a particular device. All amplifiers have an upper limit on their output power; increasing the input power further will not produce any change in the output beyond this point. This effect is called **gain saturation**. There are various types of noise sources which affect optical amplifiers; in particular, these devices are sensitive to the polarization of the input light. **Polarization sensitivity** refers to the variations in amplifier gain with changes in the signal polarization.

Optical amplifiers allow the signal to remain in optical form throughout a link or network. One advantage of this approach is that it becomes possible to change the signal data rates without affecting the amplifier, or to transmit multiple data rates on a single fiber at different wavelengths, and still achieve the desired amplification. Their main disadvantage, compared to repeaters, is that they do not retime the signal so that timing jitter and dispersion will accumulate through a long chain of optical amplifiers. There are two main types of optical amplifiers, namely fiber amplifiers, also called **rare earth doped optical amplifiers** (where the gain is provided by optical pumping of a fiber), and **semiconductor optical amplifiers** (where the gain medium is electrically pumped). It is also possible to make optical amplifiers based on nonlinear effects such as Raman or Brilloun scattering.

5.2.1 *RARE EARTH DOPED OPTICAL FIBER AMPLIFIERS*

When optical fibers are doped with rare earth ions, such as erbium (Er), neodymium (Nd) or praseodymium (Pd), they can be pumped by a diode laser so that the outer electrons of the rare earth ions are raised to higher energy levels. The resulting population inversion allows the fiber to act like a laser; if light incident on these fibers is at the proper wavelength, then the fiber will produce additional photons at the same wavelength, thus amplifying the optical signal. In this way, a diode laser of the proper wavelength acts as the source which optically pumps the doped fiber to create a population inversion, as discussed in Chapter 2. Doping the fiber with different rare earth elements results in operation at different optical wavelengths. For example, Nd doped fiber amplifies wavelengths around 1.06 microns, Pd doped fiber works around 1.3 microns, and Er doped fiber works around 1.55 microns. Since long distance communication links typically use wavelength near 1.55 microns anyway because this is the attenuation minimum of glass fiber, **erbium doped fiber amplifiers (EDFAs)** are most useful for long distance communications applications.

Erbium doped fiber amplifiers allow the amplification of optical signals along their direction of travel in a fiber, without the need to convert back and forth from the electrical domain. An EDFA operates on the same principle as an optically pumped laser; it consists of a relatively short (about 10 m) section of fiber doped with a controlled amount of Er ions. When this fiber is pumped at high power (10–300 mW) with light at the proper wavelength (either 980 or 1480 nm) the Er ions absorb the light and are excited to a higher energy state. Another incident photon around 1550 nm wavelength will cause stimulated emission of light at the same wavelength, phase, and direction of travel as the incident signal. Using these pump wavelengths, the EDFA can amplify wavelengths of light between about 1520 and 1620 nm. Other wavelengths can be used in principle to excite an EDFA (e.g. 514 nm, 532 nm, 667 nm, and 800 nm) but these are not as common since high-power pump lasers at these wavelengths are not readily available. EDFAs are often characterized by their **gain coefficient**, defined as the small signal gain divided by the pump power. As the input power is increased, the total gain of the EDFA will slowly decrease; at some point, the EDFA enters **gain saturation**, and further increases to the input power cease to result in any increase in output power. The **output saturation power** is defined as the output power level at which the gain has dropped by 3 dB. Since the EDFA does not distort the signal, unlike electronic amplifiers, they are often used in gain saturation mode. While there are other types of optical amplifiers based on other rare earth elements such as Pd or Nd, and even some optical amplifiers based on semiconductor devices, the Er doped amplifiers are the most widely used because of their maturity and good performance at wavelengths of interest near 1550 nm.

Since EDFAs do not require converting the signal to electronic form, they are well suited to long haul optical communication networks and WDM systems. It is possible to achieve high efficiencies when coupling optical power from the pump to the signal (over 50%), and to amplify most of the common WDM wavelengths in parallel with a single amplifier (dynamic range around 80 nm). Separate amplifiers are required to amplify the C-band and L-band wavelengths in a WDM network. Gain as high as 40 dB can be achieved, which is relatively constant (under 20 dB variation) across the amplifier passband. However, the more wavelengths used the less power per wavelength is available. Thus, if we drop some wavelengths from the WDM network the EDFA will amplify fewer additional wavelengths; similarly, if we add new wavelengths they will be amplified less. Since the gain does not remain the same between one add/drop multiplexing point to the next, a WDM system with EDFAs must compensate through other components (dynamic gain equalizers). Saturation output for an EDFA is usually greater than 1 mW (10 dB). To achieve the highest gains, EDFAs may be quite long, up to several km. EDFAs inherently produce some output light even when there is no input; this is called spontaneous emission noise, and degrades the SNR in the link. There are other noise sources, such as crosstalk and four-wave mixing, which are discussed in Chapter 6. Controlling optical power in a WDM system is a complex problem; power levels must be kept between a lower and a upper limit to trade off gain with the various noise sources in a link.

Figure 5.1 shows a typical configuration for an optical amplifier. The fiber is positioned between polarization independent optical isolators to minimize stray light feedback into the amplifier. Pump light from the diode laser is input by a wavelength dependent coupler so that only the desired pump wavelength passes through. The length of the fiber is

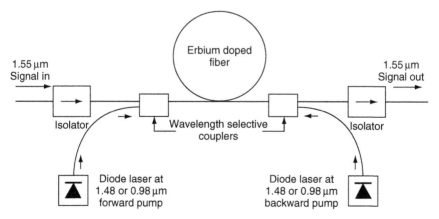

Figure 5.1   General erbium doped fiber configuration showing bidirectional pumping.

chosen to optimize gain; if it is too long, some re-absorption of the signal will occur, but if it is too short the amplifier will not make full use of the available pump energy.

The gain of a typical EDFA varies strongly with wavelength; for example, the gain at 1560 nm is about twice as large as the gain at 1540 nm. This can be a problem when operating WDM systems; some channels will be strongly amplified and dominate over other channels that are lost in the noise. Furthermore, a significant complication with EDFAs is that their gain profile changes with input signal power levels; so, for example, in a WDM system the amplifier response may become nonuniform (different channels have different effective gain) when channels are added or dropped from the fiber. This requires some form of **equalization** to achieve a flat gain across all channels. There has been a great deal of research in this area; some proposals include adding an extra WDM channel locally at the EDFA to absorb excess power (**gain clamping**), and manipulating either the fiber doping or the core structure. Another concern with EDFAs is that some of the excited Er undergoes spontaneous emission, which can create light propagating in the same direction as the desired signal. This random light is amplified and acts as background noise on the fiber link; the effect is known as **ASE**. Since ASE can be at the same wavelength as the desired signal, it may be difficult to filter out; furthermore, ASE accumulates in systems with multiple amplifier stages and is proportional to the amplifier gain.

In the wavelength region around 1280–1340 nm, where EDFAs do not function, it is possible to use praseodymium-doped fiber amplifiers (PDFAs) to achieve fairly high gain (around 30 dB). However, such devices require a special fiber made of fluoride rather than silica glass, and a high power pump (over 300 mW) at a wavelength of 1017 nm; neither the fiber nor the pump lasers are widely available. Similar problems exist for fiber amplifiers doped with different materials or intended for other wavelengths, such as Nd doped fibers or fibers which contain elements such as gallium-lanthanium-sulfide and gallium-lanthanium-iodine.

Fiber amplifiers can also be built by taking advantage of nonlinear optical effects, including **Raman** and **Brillouin** third-order nonlinearities in the glass. These effects are related to interactions between photons and acoustical phonons created from vibrations of the molecular lattice within the core of the fiber, and will be discussed later in this chapter. Brillouin amplification is mostly used for receiver pre-amplification and for moderate bit rate communication systems (<100 MHz). The Raman amplification bandwidth is very large (>5 THz), which implies that Raman amplifiers can be used for very high-bit-rate data systems, for any wavelength region, being limited only by the available pump sources.

5.2.2 SEMICONDUCTOR AMPLIFIERS

A semiconductor amplifier is similar to a laser diode operating below its laser threshold. There are two basic types: Fabry-Perot amplifiers (FPA) and traveling wave amplifiers (TWA). The main difference is in the amount of light which is reflected back into the amplification media from the attached mirror facets; facet reflectivities are approximately 0.3 for a FPA and 10^{-3}–10^{-4} for a TWA. A schematic of a SLA is shown in Figure 5.2, along with a graph of how light output is related to pump current and facet reflectivity.

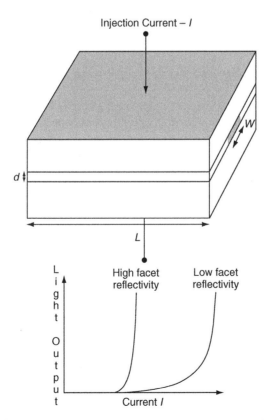

Figure 5.2 Schematic of an SLA and its light output vs. injection current characteristics for two different facet reflectivities.

The maximum available signal is limited by gain saturation. In a TWA, the gain G_s can be calculated by

$$G_s = 1 + (P_{sat}/P_{in}) \ln (G_o/G_s) \qquad (5.1)$$

Where G_o is the maximum amplifier gain and P_{sat}/P_{in} is the ratio of the saturation power and the input power.

Cross talk can occur when multiple optical signals or channels are amplified simultaneously. It is most pronounced when the data rate is comparable to the reciprocal of the carrier lifetime.

A source of noise in these devices is ASE, or the amplification of stray light produced by random photon emissions in the material. This is to be expected for devices whose design is similar to a laser with feedback mechanisms. This noise produces a mixed or beat frequency with the signal, causing additional noise sources known as signal-spontaneous beat noise and spontaneous-spontaneous beat noise. Minimization of these noise sources is a difficult problem in amplifier design.

5.2.3 NONLINEAR EFFECTS AND OPTICAL AMPLIFIERS FOR WDM

The interaction of light with the optical fiber material is typically very small, particularly at low optical power levels. However, as the level of optical power in the fiber is increased nonlinear effects can become significant; this is especially important for long distance fiber links with amplification, since they provide the opportunity for smaller effects to build up over distance. Since WDM involves transmitting many optical signals over a common fiber, nonlinear effects in the link can become significant here as well. In an optical communication system, nonlinear effects can induce transmission errors which place fundamental limits on system performance, in much the same fashion as attenuation or dispersion effects. At the same time, some important components such as amplifiers for extended distance WDM systems rely on nonlinear effects for their operation [1]. There is a design tradeoff in the appropriate use of these effects within an optical communication network.

One of the most common nonlinear interactions is known as "**four wave mixing**" (**FWM**), which occurs when two or more optical signals propagate in the same direction along a common single-mode fiber. Optical signals in the fiber can mix to produce new signals at wavelengths which are spaced at the same intervals as the original signals. The effect can also occur between three or more signals, making the overall effect quite complex. FWM increases exponentially with signal power, and becomes greater as the channel spacing is reduced; in particular, it is a concern with dense WDM systems. If WDM channels are evenly spaced, then the spurious FWM signals will appear in adjacent wavelength channels and act as noise. One method of dealing with this problem is to space the channels unevenly to reduce the effect of added noise on adjacent channels; however, FWM still removes some optical power from the desired signal levels. Since FWM is caused by signals which remain in phase with each other over a significant propagation distance, the effect is stronger for lasers with a long coherence length (i.e. they remain coherent with each other over a significant propagation distance before starting to drift out of phase). Also, FWM is strongly influenced by chromatic dispersion; because dispersion insures that different signals do not stay in phase with each other for very long, it acts to reduce the effect of FWM.

Gain saturation occurs in an optical amplifier when high power is injected into the amplifier gain media. If two wavelength channels are passing through an amplifier, one of which is modulated with data and the other is not (continuous wave operation), then crosstalk may occur between the two wavelengths. When the modulated data stream has high power (logical "1"), depletion in the amplifier gain medium will tend to remove power from the second wavelength. Likewise, when the modulate data stream has low power (logical "0"), depletion does not occur and the amplifier will pass through power on the second wavelength. Thus, the second wavelength picks up an inverted copy of the data on the first wavelength. This effect is called **cross-gain modulation**, and variations on this can cause problems in the design of practical optical amplification systems.

At higher optical power levels, nonlinear scattering may limit the behavior of a fiber optic link. The dominant effects are stimulated Raman and Brillouin scattering. When incident optical power exceeds a threshold value, significant amounts of light may be scattered from small imperfections in the fiber core or by mechanical (acoustic) vibrations in the transmission media. These vibrations can be caused by the high intensity electromagnetic fields of light concentrated in the core of a single-mode fiber. Because the scattering process also involves the generation of photons, the scattered light can be frequency shifted [2–5]. Put another way, we can think of the high intensity light as generating a regular pattern of very slight differences in the fiber refractive index; this creates a moving diffraction grating in the fiber core, and the scattered light from this grating is Doppler shifted in frequency by about 11 GHz. This effect is known as **stimulated Brillouin scattering (SBS)**; under these conditions, the output light intensity becomes nonlinear as well. SBS will not occur below a critical optical power threshold, P, given by [1]

$$P = \frac{21A}{GL} \text{ watts} \tag{5.2}$$

where A is the cross-section area of the fiber for the guided mode being amplified, G is a material property called the Brillouin gain coefficient, and L is the fiber length. Brillouin scattering has been observed in single-mode fibers at wavelengths greater than cutoff with optical power as low as 5 mW; it can be a serious problem in long distance communication systems when the span between amplifiers is low and the bit rate is less than about 2 Gbit/s, in WDM systems up to about 10 Gbit/s when the spectral width of the signal is very narrow, or in remote pumping of some types of optical amplifiers. In general, SBS is worse for narrow laser linewidths (and is generally not a problem for channel bandwidth greater than 100 MHz), wavelengths used in WDM (SBS is worst near 1550 nm than near 1300 nm), and higher signal power per unit area in the fiber core. SBS can be a concern in long distance communication systems, when the span between amplifiers is large and the bit rate is below about 2.5 Gbit/s, in WDM systems where the spectral width of the signal is very narrow, and in remote pumping of optical amplifiers using narrow linewidth sources. In cases where SBS could be a problem, the source linewidth can be intentionally broadened by using an external modulator or additional RF modulation on the laser injection current. However, this is a tradeoff against long distance transmission, since broadening the linewidth also increases the effects of chromatic dispersion.

When the scattered light experiences frequency shifts outside the acoustic phonon range, due instead to modulation by impurities or molecular vibrations in the fiber core, the effect is known as **stimulated Raman scattering (SRS)**. The mechanism is similar to SBS, and scattered light can occur in both the forward and the backward directions along the fiber; the effect will not occur below a threshold optical power level given by [1]

$$P = \frac{16A}{GrL} \text{ watts} \tag{5.3}$$

where A and L are the same as in the case of SBS, and Gr is a constant called the Raman gain coefficient. As a rule of thumb, the optical power threshold for Raman scattering is about three times larger than for Brillouin scattering. Another good rule of thumb is that SRS can be kept to acceptable levels if the product of total power and total optical bandwidth is less than 500 GHz-W. This is quite a lot; for example, consider a 10-channel DWDM system with standard wavelength spacing of 1.6 nm (200 GHz). The bandwidth becomes $200 \times 10 = 2000$ GHz, so the total power in all 10 channels would be limited to 250 mW in this case (in most DWDM systems, each channel will be well below 10 mW for other reasons such as laser safety considerations).

In single-mode fiber, typical thresholds for Brillouin scattering are about 10 mW and for Raman scattering about 35 mW; these effects rarely occur in multimode fiber, where the thresholds are about 150 mW and 450 mW, respectively. In general, the effect of SRS becomes greater as the signals are moved further apart in wavelength (within some limits); this introduces a tradeoff with FWM, which is reduced as the signal spacing increases.

Optical amplifiers can also be constructed using the principle of SRS; if a pump signal with relatively high power (0.5–1 watt or more) and a frequency 13.2 THz higher than the signal frequency is coupled into a sufficiently long length of fiber (greater than 1 km), then amplification of the signal will occur. Unfortunately, more efficient amplifiers require that the signal and pump wavelengths be spaced by almost exactly the **Raman shift** of 13.2 THz, otherwise the amplification effect is greatly reduced. It is not possible to build high power lasers at arbitrary signal wavelengths; one possible solution is to build a pump laser at a convenient wavelength, then wavelength shift the signal by the desired amount. Typically, a Raman amplifier will have its pump laser located near the receive end of the link, and the pump light will travel back along the fiber towards the signal source. This has the advantage of making the pump power strongest where it is most needed (far away from the signal source); the signal is amplified most when it is weakest and less where it is strongest. Other, experimental arrangements have been proposed which use a pump at both ends of the fiber to simultaneously amplify the C-band and L-band wavelengths. Raman amplifiers offer a wide bandwidth (1300 nm to over 1600 nm, better than 500 channels at 100 GHz spacing). They also have no restrictions on gain over their bandwidth, which enables multiterbit transmission techniques often called the **Raman supercontinuum**. However, they also face safety and thermal management issues, since the pump lasers have high optical power.

References

[1] The following ITU standards deal with optical amplifiers:
  - EDFAs, line amps, booster amps, and optical amplifier subsystems are described in G.622.
  - Remotely pumped amplifiers are described in G.973.
[2] Miller and Chylworth (1985) *Optical Communications*, Prentice Hall.
[3] DeCusatis, C. (ed.) (2002) *Handbook of Fiber Optic Data Communications* 2nd edition, Academic Press.
[4] Bass, M. (ed.) (2001) *Handbook of Optics*, vol. III, McGraw-Hill.
[5] Dutton, H. (1999) *Understanding Optical Communications*, IBM Press.

Chapter 6 | Wavelength Multiplexing

Multiplexing wavelengths is a way to take advantage of the high bandwidth of fiber optic cables without requiring extremely high modulation rates at the transceiver. Normally, one communication channel requires two optical fibers, one to transmit and the other to receive data; a multiplexer provides the means to run many independent data or voice channels over a single pair of fibers. This device takes advantage of the fact that different wavelengths of light will not interfere with each other when they are carried over the same optical fiber; this principle is known as WDM. The concept is similar to frequency multiplexing used by FM radio, except that the carrier "frequencies" are in the optical portion of the spectrum (around 1550 nm wavelength, or 2×10^{14} Hz). Thus, by placing each data channel on a different wavelength (frequency) of light, it is possible to send many channels of data over the same fiber. More data channels can be carried if the wavelengths are spaced closer together; this is known as dense wavelength division multiplexing (DWDM). The available optical spectrum is often divided into several different wavelength bands, as illustrated in Table 6.1. Following international standards, the wavelength spacing for our product is 1.6 nm, or about 200 GHz. The product is protocol independent; it provides a selection of fiber optic interfaces to attach any type of voice or data communication channel. Input data channels are converted from optical to electrical signals, routed to an appropriate output port, converted into optical DWDM signals, and then combined into a single channel (this is done using optical filters which selectively pass or block a narrow range of wavelengths). The combined wavelengths are carried over a single pair of fibers; another multiplexer at

Table 6.1   **Optical wavelength bands**

Original (O-band)	1260–1360 nm
Extended (E-band)	1360–1460 nm
Short (S-band)	1460–1530 nm
Conventional (C-band)	1530–1565 nm
Long (L-band)	1565–1625 nm
Ultralong (UL-band)	1625–1675 nm

the far end of the fiber link reverses the process and provides the original data streams. The product also retimes the signals to reduce noise, and increases the transmission distance up to 200 km, far beyond the normal limits of the input protocols. This product is the first to offer support for point-to-point, hubbed ring, and meshed ring networks. It can be used to add or drop data channels at any point on a network, and wavelengths may be reused to increase the network's capacity to over 2000 channels. The network is fault tolerant and self-healing with protection switching; if a fiber path between two locations breaks, the remaining intact locations continue to operate uninterrupted. The entire network can be managed from one personal computer which may be hundreds of kilometers away from the multiplexers.

With an available bandwidth of about 25 THz, a single optical fiber could carry all the telephone traffic in the U.S. on the busiest day of the year (recently, Mother's Day has been slightly exceeded by Valentine's Day). This technology represents an estimated $1.6 billion market with over 50% annual growth; wavelength multiplexing systems may be classified according to their wavelength spacing and number of channels as follows [1].

- Coarse WDM systems (CWDM) were first developed with only 2–3 wavelengths widely spaced, for example 1300 nm and 1550 nm. This may be useful for networks that require a low cost solution for bi-directional communication on a single fiber, such as some types of cable television. Subsequently, CWDM systems with 4, 8, or 16 channels were developed for use in small networks and as a supplement to more dense wavelength multiplexing schemes. This approach has also been used within optical transceivers to achieve higher data rates, for example, by combining four wavelengths of data at 2.5 Gbit/s each to achieve an aggregate 10 Gbit/s over a single fiber.

- Wide spectrum WDM (WWDM) systems can support up to 16 channels, using wavelengths which are spaced relatively far apart; there is no standardized wavelength spacing currently defined for such systems, although spacing of 1 to 30 nm have been employed. They are meant to serve as a low cost alternative to DWDM for applications which do not require large numbers of channels on a single fiber path, and are being considered as an option for the emerging 10 Gbit/s Ethernet standard. These are also known as sparse WDM (SWDM) systems.

- DWDM employed wavelengths spaced much closer together, typically following multiples of the International Telecommunications Union (ITU) industry standard grid [2] with wavelengths near 1550 nm and a minimum wavelength spacing of 0.8 nm (100 GHz). This may be further subdivided as follows:

  - First generation DWDM systems typically employed up to eight full duplex channels multiplexed into a single duplex channel.

  - Second generation DWDM systems employ up to 16 channels.

  - Third generation DWDM systems employ up to 32 channels; this is the largest system currently in commercial production for data communication applications.

  - Fourth generation or Ultra-dense WDM are expected to employ 40 channels or more and may deviate from the current ITU grid wavelengths; channel spacing as small as 0.4 nm (50 GHz) have been proposed [3]. These systems are not yet commercially available, and are the subject of much ongoing research.

Most CWDM systems today have standardized around 20 nm wavelength spacing; the most commonly used schemes include four wavelengths in the 1510–1570 nm band, eight wavelengths in the 1470–1610 nm band, and 16 wavelengths in the 1310–1610 nm band. Most DWDM development has centered around the C-band and L-band wavelengths. Following standards set by the ITU, the wavelength spacing for DWDM products is a minimum of 0.8 nm, or about 100 GHz; in practice, many products use a slightly broader spacing such as 1.6 nm, or about 200 GHz, to simplify the design and lower overall product cost. In addition to the distinctions between C-band and L-band wavelengths shown here, some Japanese standards have begun to define wavelengths in the

S-band between 1470 and 1520 nm; however, these are not widely used at present. Currently, it is less expensive to implement C-band wavelengths than L-band for various technical reasons which will be discussed in later chapters. This is usually more than adequate for current applications, so it is not likely that there will be a large amount of multiplexing deployed outside of the C and L bands in the near future. There are various technical problems associated with working outside of these two bands; for example, much of the installed fiber infrastructure has high attenuation in the E, O, and UL bands (the E band attenuation can be overcome to some degree by using hydroxyl reduced fiber). The concept of combining multiple data channels over a common fiber (physical media) is illustrated in Figure 6.1. Note that the process is in principle protocol independent; it provides a selection of fiber optic interfaces to attach any type of voice or data communication channel. Input data channels are converted from optical to electrical signals, routed to an appropriate output port, converted into optical DWDM signals, and then combined into a single channel. The wavelengths may be combined in many ways; for example, a diffraction grating or prism may be used. Both of these components act as dispersive optical elements for the wavelengths of interest; they can separate or recombine different wavelengths of light. The prism or grating can be packaged with fiber optic pigtails and integrated optical lenses to focus the light from multiple optical fibers into a single optical fiber; demultiplexing reverses the process. The grating or prism can be quite small, and may be suitable for integration within a CWDM transceiver package. Such components are typically fabricated from a glass material with low coefficient of thermal expansion, since the diffraction properties change with temperature. For CWDM applications, this is not an issue because of the relatively wide spacing between wavelengths. For DWDM systems, the optical components must often be temperature compensated

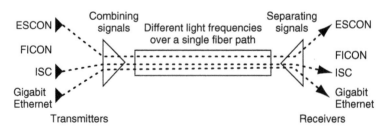

Figure 6.1 Dense Wavelength division multiplexing (DWDM).

using heat sinks or thermoelectric coolers to maintain good wavelength separation over a wide range of ambient operating temperatures.

Because of the precision tolerances required to fabricate these parts and the complex systems to keep them protected and temperature stable, this solution can become expensive, especially in a network with many add/drop locations (at least one grating or prism would be required for each add/drop location). Another form of multiplexing element is the array waveguide grating (AWG); this is basically just a group of optical fibers or waveguides of slightly different lengths, which controls the optical delay of different wavelengths and thereby acts like a diffraction grating. AWGs can be fabricated as integrated optical devices either in glass or silicon substrates, which are thermally stable and offer proven reliability. These devices also provide low insertion loss, low polarization sensitivity, narrow, accurate wavelength channel spacing, and do not require hermetic packaging; one example of a commercial product is the Lucent Lightby 40 channel AWG mux/demux. Recently, the technology for fabricating in-fiber Bragg gratings as an alternative approach has shown promise; this will be discussed later in the chapter. Another promising new technology announced recently by Mitel Corporation's semiconductor division implements an Echelle grating by etching deep, vertical grooves into a silica substrate; this process is not only compatible with standard silicon processing technology, which may offer the ability to integrate it with other semiconductor WDM components, but also offers a footprint for a 40-channel grating up to five times smaller than most commercially available devices.

However, in many commercially available DWDM devices, a more common approach is to use thin film-interference filters on a glass or other transparent substrate. These multilayer filters can selectively pass or block a narrow range of wavelengths; by pig-tailing optical fibers to the filters, it is possible to either combine many wavelengths into a single channel or split apart individual wavelengths from a common fiber. Note that many filters may be required to accommodate a system with a large number of wavelengths, and each filter has some insertion and absorption loss associated with it; this can affect the link budget in a large WDM network. For example, a typical four-channel thin film add/drop filter can have as much as 3–4 dB loss for wavelengths which are not added or dropped; a cascade of many such filters can reduce the effective link budget and distance of a network by 10–20 km or more. The combined wavelengths are carried over a single pair of fibers; another multiplexer at the far end of the fiber link reverses the process and provides the original data streams.

6.1   WDM Design Considerations

There are many important characteristics to consider when designing a DWDM system. One of the most obvious design points is the largest total number of channels (largest total amount of data) supported over the multiplexed fiber optic network. Typically one wavelength is required to support a data stream; duplex data streams may require two different wavelengths in each direction or may use the same wavelength for bi-directional transmission. As we will discuss shortly, additional channel capacity may be added to the network using a combination of WDM and other features, including time division multiplexing (TDM) and wave-length reuse. Wavelength reuse refers to the product's ability to reuse the same wavelength channel for communication between multiple loca-tions; this increases the number of channels in the network. For example, consider a ring with three different locations A, B, and C. Without wave-length reuse, the network would require one wavelength to communicate between sites A and C, and another wavelength to communicate between sites B and C, or two wavelength channels total. With wavelength reuse, the first and second sites may communicate over one wavelength, then the second and third sites may reuse the same wavelength to communicate rather than requiring a new wavelength. Thus, the second wavelength is now available to carry other traffic in the network. Wavelength reuse is desirable because it allows the product to increase the number of physical locations or data channels supported on a ring without increasing the number of wavelengths required; a tradeoff is that systems with wave-length reuse cannot offer protection switching on the reused channels. TDM is another way in which some WDM products increase the number of channels on the network. Multiple data streams share a common fiber path by dividing it into time slots, which are then interleaved onto the fiber. TDM acts as a front-end for WDM by combining several low data rate channels into a higher data rate channel; since the higher data rate channel only requires a single wavelength of the WDM, this method provides for increased numbers of low speed channels. As an example, if the maximum data rate on a WDM channel is 1 Gbit/s then it should be possible to TDM up to four channels over this wavelength, each with a bit rate of 200 Mbit/s, and still have some margin for channel over-head and other features. This is also sometimes known as wavefill, or sub-rate multiplexing (SRM). The TDM function may be offered as part of a separate product, such as a data switch, that interoperates with the wavelength multiplexer; preferably, it would be integrated into the WDM

Table 6.2 Common SONET networking hierarchy

Signal	Bit rate	Voice slots
DS0	64 kbps	1 DS0
DS1	1.544 Mbps	24 DS0
DS2	6.312 Mbps	96 DS0
DS3	44.736 Mbps	28 DS0

design. Some products only support TDM for selected telecommunications protocols such as SONET; in fact, this telecom protocol is designed to function in a TDM-only network, and can easily be concatenated at successively faster data rates. The basic data rates and notation associated with SONET networks are shown in Table 6.2.

Note that the original unit used in multiplexing voice traffic is 64 kbps, which represents one phone call. Within North America, 24 of these units are TDM multiplexed into a signal with an aggregate speed of 1.544 Mbit/s for transmission over T1 lines. Outside of North America, 32 signals are TDM multiplexed into a 2.048 Mbit/s signal for transmission over E1 lines.

Optical carrier	SONET/SDH signal	Bit rate	Capacity
OC-1	STS-1	51.84 Mbps	28 DS1s or 1 DS3
OC-3	STS-3/STM-1	155.54 Mbps	84 DS1 or 3 DS3
OC-12	STS-12/STM-4	622.08 Mbps	336 DS1 or 12 DS3
OC-48	STS-48/STM-16	2488.32 Mbps	1344 DS1 or 48 DS3
OC-192	STS-192/STM-64	9953.28 Mbps	5379 DS1 or 192 DS3

Using virtual tributaries, 28 DS1 signals can be mapped into the STS-1 payload, or multiplexed to DS3 with an M13 multiplexer and fit directly into STS-1.

However, since WDM technology can be made protocol independent, it is desirable for the TDM to also be bit rate and protocol independent, or at least be able to accommodate other than SONET-based protocols. This is sometimes referred to as being "frequency agile." Note that while a pure TDM network requires that the maximum bit rate continue to increase in order to support more traffic, a WDM network does not require the individual channel bit rates to increase. Depending on the type of network being used, a hybrid TDM and WDM solution may offer the best overall cost performance; however, TDM alone does not scale as well as WDM.

Another way to measure the capacity of the multiplexer is by its maximum bandwidth, which refers to the product of the maximum number of channels and the maximum data rate per channel. For example, a product which supports up to 16 channels, each with a maximum data rate of 1.25 Gigabit per second (Gbps) has a maximum bandwidth of 20 Gbps. Note that the best way to measure bandwidth is in terms of the protocol independent channels supported on the device; some multiplexers may offer very large bandwidth, but only when carrying well-behaved protocols such as SONET or SDH, not when fully configured with a mixture of datacom and telecom protocol adapters. Another way to measure the multiplexer's performance is in terms of the maximum number of protocol independent, full duplex wavelength channels which can be reduced to a single channel using wavelength multiplexing only. Some products offer either greater or fewer numbers of channels when used with options such as a fiber optic switch.

Another important consideration in the design of WDM equipment is the number of multiplexing stages (or cards) required. It is desirable to have the smallest number of cards supporting a full range of datacom and telecom protocols. Generally speaking, a WDM device contains two optical interfaces, one for attachment of input or client signals (which may be protocol specific) and one for attachment of the WDM signals. Each client interface may require a unique adapter card; for example, some protocols require a physical layer which is based on an LED transmitter operating over 62.5 micron multimode fiber, others use short wavelength lasers with 50 micron multimode fiber, and still others require long wavelength lasers with single-mode fiber. Likewise, each channel on the WDM interface uses a different wavelength laser transmitter tuned to an ITU grid wavelength, and therefore requires a unique adapter card. Some designs place these two interfaces on a single card, which means that more cards are required to support the system; as an example, a product with 16 wavelength channels may require 16 cards to support ESCON, 16 more to support ATM, and in general to support N channels with M protocols would require $N \times M$ cards. Typically $N = 16$ to 32 channels and $M = 10$ to 15 protocols, so this translates into greater total cost for a large system, greater cost in tracking more part numbers and carrying more spare cards in inventory, and possibly lower reliability (since the card with both features can be quite complex). This common card is sometimes known as a transponder, especially if it offers only optical input and output interfaces. An advantage of the transponder design is that all connections to the product are made with optical fiber; the backplane does not carry high speed signals, and upgrades to the system can be made more easily by swapping adapter cards. However, a cross-connected high bandwidth backplane is still a desirable feature to avoid backplane bus congestion at higher data rates and larger channel counts; furthermore, it provides the possibility of extending the backplane into rack-to-rack type interconnections using parallel optical interconnects. Another possible design point places the client interface and WDM interface on separate cards, and uses a common card on the client side to support many protocol types, so only one card for each protocol is required. Continuing our example, a product with 16 channels may support 10 different protocols; using the first design point discussed above this requires 160 cards, while the second design point requires only 16 cards, a significant simplification and cost savings. The tradeoff with this second design point is that more total cards may be required to populate the system, since two cards are required for every transponder, or in other words a larger footprint for the same number of wavelength channels. In practice, a common client interface card may be configurable to support many different protocols by using optical adapters and attenuators at the interface, and making features such as retiming programmable for different data rates. This means that significantly less than one card per protocol is required; a maximum of 2–5 cards should be able to support the full range of networking protocols. Note that WDM devices which do not offer native attachment of all protocols may require separate optical patch panels, strain relief, or other protection for the optical fibers; this may require additional installation space and cost. Features such as adapters or patch panels which must be field installed also implies that a particular configuration cannot be fully tested before it reaches the end user location; field installed components also tend to have lower reliability than factory build and installed components. For example, if a product supports ESCON protocols, a user should be able to plug an ESCON duplex connector directly into the product as delivered, without requiring adapters for the optical connectors. Note that some *ad hoc* industry standards such as the Parallel Sysplex architecture for coupling mainframe computers have very specific performance requirements, and may not be supported on all DWDM platforms; recently published technical data from a refereed journal is a good reference to determine which protocols have been tested in a given application; there may also be other network design considerations, such as configuring a total network solution in which the properties of the subtended equipment are as well understood as those of the DWDM solution.

There are several emerging technologies which may help address these design points in the near future. One example is wavelength tunable or agile lasers, whose output wavelength can be adjusted to cover several possible wavelengths on the ITU grid. This would mean that fewer long wavelength adapter cards or transponders would be required; even if the wavelength agile laser could be tuned over only two ITU grid wavelengths, this would still cut the total number of cards in half. There are basically three ways to tune a laser wavelength, all of which rely on changing the optical path length of the laser diode cavity. These include mechanical control of a mirror or grating, which may incorporate microelectromechanical devices (MEMs), temperature control, or varying the laser diode drive current. Tunable lasers are available commercially from various sources, including Fujitsu and Lucent; new tuning methods are also being investigated. Another potentially useful technology is pluggable optical transceivers, such as the gigabit interface converter (GBIC) package or emerging pluggable small form factor (SFF) transceivers. This could mean that an adapter card would be upgradable in the field, or could be more easily repaired by simply changing the optical interface; this reduces the requirement to keep large numbers of cards as field spares in case of failure. Also, a pluggable interface may be able to support the full physical layer of some protocols, including the maximum distance, without the need for optical attenuators, patch panels, or other connections; the tradeoff for this native attachment is that changing the card protocol would require swapping the optical transceiver on the adapter card.

Using currently available technology, it is much easier to implement ITU grid lasers and optical amplifiers at conventional or C-band wavelengths than at emerging long band or L-band wavelengths. Hence, some implementations use tighter wavelength spacing in order to fit 32 or more channels in C-band; the scalability of this approach remains open to question. Other approaches available today make use of L-band wavelength and larger inter-wavelength spacing, and can be more easily scaled to larger wavelength counts in the future. Performance of the multiplexer's optical transfer function (OTF), or characterization of the allowable optical power ripple as a function of wavelength, is an important parameter in both C-band and L-band systems.

6.2   Network Topologies

Modern optical communication networks are in a state of ongoing evolution and change. While the distinction between different parts of the
network is not always clear-cut, one simplified way to view the global
network topology is a hierarchical view [4]. Long haul networks form the
core of the global network, typically based on SONET/SDH protocols.
These networks are controlled by a relatively small group of transnational
carriers, and their primary technical concern is providing sufficient transport bandwidth to meet demands. At the other extreme are the access
networks, which are closest to the end users (ranging from individual residential users to large corporations and institutions). These networks are
characterized by diverse protocols and infrastructures, and are controlled
by a wide range of companies; they face the challenge of incorporating
traditional voice traffic with data traffic (such as ESCON, Fibre Channel,
or others) and the explosion of IP traffic, which is inherently bursty,
asymmetric, and unpredictable. In the United States, competition was
introduced into the long distance phone market in 1984, and the 1996
Telecommunications Reform Act has resulted in a broad range of new
carriers. Between these two large domains lies the metropolitan area
network (MAN), which channel traffic between long haul points of presence (POP), and within the metropolitan domain (perhaps a 100 km or
more). This network is a critical element, which must balance the ever-
increasing bandwidth demand of long haul networks with the flexibility
and diversity of the access network. While there is a natural tendency
to regard the MAN as simply a scaled-down version of the long haul
network, this can be very misleading. Network topologies change frequently in the MAN, and must content with a much broader range of
protocol options; furthermore, they include a collection of lower bit rate
services, both synchronous and asynchronous, short loops, small network
cross-sections, and a wide range of bandwidth demands. Thus, while
capacity is the overriding requirement in the long haul market, the MAN
is characterized by a more diverse set of requirements including protocol
and data rate transparency, scalability, and dynamic provisioning. Other
views define the access network as simply a part of the MAN, and further breaks down the metropolitan market segments into the core (scaled
down long haul systems which interconnect carrier POPs), the enterprise
(large corporate users with leased fiber that serve storage area networks
and geographically distributed applications such as disaster recovery)
and the metro access market (the segment between carrier POPs and

access facilities, including both traffic aggregation points and customer premises).

Conventional SONET networks are designed for the WAN and are based on reconfigurable ring topologies, while most datacom networks function as switched networks in the LAN and point-to-point in the WAN or MAN. There has been a great deal of work done on optimizing nation-wide WANs for performance and scalability, and interfacing them with suitable LAN and MAN topologies; WDM plays a key role at all three network levels, and various traffic engineering approaches will be discussed later in this chapter. Despite the protocol independent nature of WDM technology, many WDM products targeted at telecom carriers or local exchange carriers (LECs) were designed to carry only SONET or SDH compatible traffic. Prior to the introduction of WDM, it was not possible to run other protocols over a ring unless they were compatible with SONET frames; WDM has made it possible to construct new types of protocol independent network topologies. For the first time, datacom protocols such as ESCON may be configured into WDM rings, including hubbed rings (a central node communicating with multiple remote nodes, also known as hub and spoke), dual hubbed rings (the same as a single hub ring except that the hub is mirrored into another backup location), meshed rings (any-to-any or peer-to-peer communication between nodes on a ring), and linear optical add/drop multiplexing (OADM) or the so-called "opened rings" (point-to-point systems with add/drop of channels at intermediate points along the link in addition to the endpoints). Some of these topologies are illustrated in Figure 6.2; we will comment specifically on a few of these options. In a hubbed ring, all channels originate and terminate on the hub node; other nodes on the ring, sometimes called satellite nodes, add and drop one or more channels which terminate at that node. Channels which are not being dropped (express channels) are optically passed through intermediate nodes, without being electronically terminated. A meshed ring refers to a physical ring that has the logical characteristics of a mesh (sometimes called a logical mesh). This is often easier to implement than a true physical mesh, which would require interconnections from every node to every other node. Note that since the DWDM network is protocol independent, care must be taken to construct networks which are functionally compatible with the attached equipment; as an example, it is possible to build a DWDM ring with attached Fibre Channel equipment that does not comply with recommended configurations such as Fibre Channel Arbitrated Loop. Other network implementations are also possible; for example, some metro

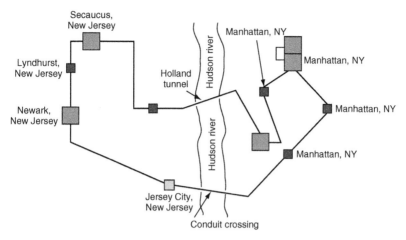

Figure 6.2   Example of a typical fiber optic ring architecture.

WDM equipment offers a two-tier ring consisting of a dual fiber ring and a separate, dedicated fiber link between each node on the ring to facilitate network management and configuration flexibility (also known as a "dual homing" architecture).

For example, Figure 6.3 depicts a typical example of a physical network in a ring topology, running from New York City across the Hudson River into northern New Jersey. There are multiple access points, or nodes, on this ring which are available for optical add/drop of network traffic. The distances involved are on the order of 100 km or less for the total ring circumference. This network is used for applications such as disaster recovery and remote backup of data at large financial institutions. The fiber paths are not straight lines because they must comply with various practical constraints, such as legal and zoning permits, right-of-way ownership, and access across the river through a limited number of available bridges or tunnels. Constraints such as these can make it difficult to identify physically redundant paths, which are desirable to protect against single points of failure in the optical network. Other examples of wide area or long haul WDM networks include the government owned Advanced Research Projects Association's ARPA-2 network, which has 21 nodes and 26 duplex interconnections in a meshed topology. Another example is the Abilene network, illustrated in Figure 6.4, which was announced by the US government in April 1998 and which serves as a precursor to the next generation optical Internet and National LambdaRail programs.

Government supported networks like these are often used as examples in the technical literature to illustrate network traffic management models.

6.3   Latency

Dense wavelength division multiplexing devices also function as channel extenders, allowing many datacom protocols to reach previously impossible distances (50–100 km or more). Combined with optical amplifier

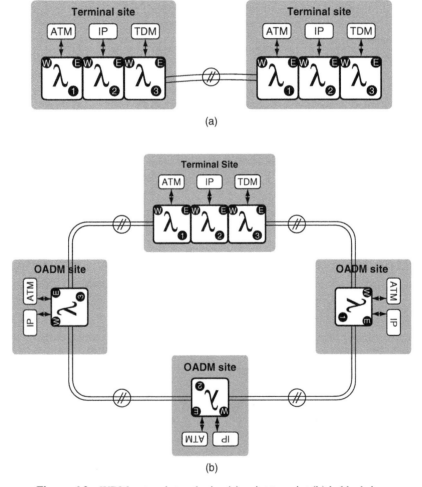

Figure 6.3   WDM network topologies (a) point-to-point (b) hubbed ring.

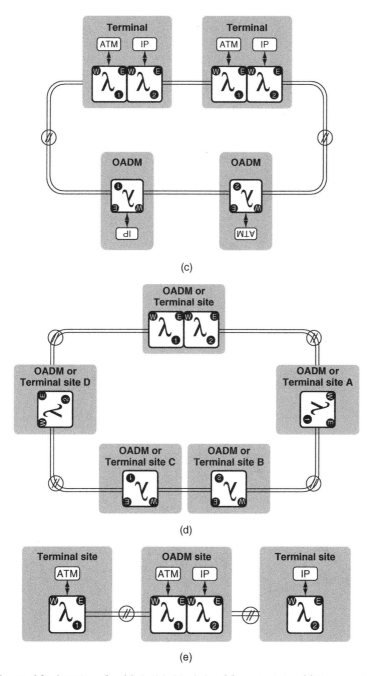

Figure 6.3 (*continued*)   (c) dual hubbed ring (d) meshed ring (e) linear optical
add/drop multiplexer (OADM).

Abilene International Network Peers

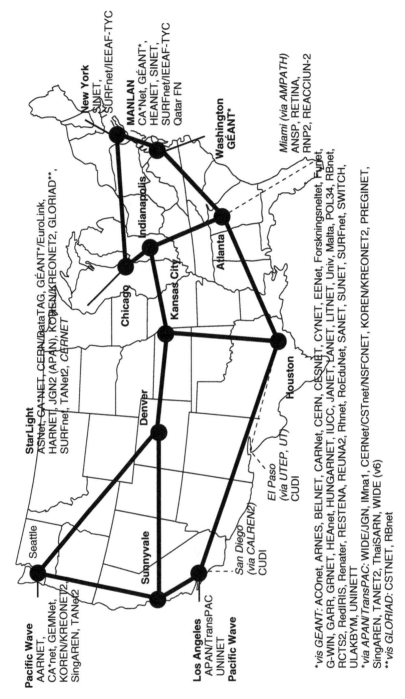

Figure 6.4 Peer nodes for the initial implementation of the next generation optical Internet.

technology, this has led some industry analysts to proclaim "the death of distance," that is connection distances should no longer pose a serious limitation in optical network design. However, in many real world applications, it is not sufficient to simply extend a physical connection; performance of the attached datacom equipment must also be considered. Latency, or propagation delay due to extended distances, remains a formidable problem for optical data communication. The effects of latency are often protocol specific or device specific. For example, using DWDM technology it is possible to extend an ESCON channel to well over 50 km. However, many ESCON control units and DASD are synchronous, and exhibit timing problems at distances beyond about 43 km. Some types of asynchronous DASD overcome this limitation; however, performance of the ESCON protocol also degrades with distance. Due to factors such as the buffer size on an ESCON channel interface card and the relatively large number of acknowledgments (ACKs) or handshakes required to complete a data block transfer (up to six or more), ESCON begins to exhibit performance droop at around 9 km on a typical channel, which grows progressively worse at longer distances. For example, at 23 km a typical ESCON channel has degraded from a maximum throughput of 17.5 MB/s to about 10 MB/s; if the application is a SAN trying to back up a petabyte database, the time required to complete a full backup operation increases significantly. This problem can be addressed to some degree by using more channels (possibly driving the need for a multiplexer to avoid fiber exhaust) or larger data block sizes, and is also somewhat application dependent. Other protocols, such as FICON, can be designed to perform much better at extended distances.

A further consideration is the performance of the attached computer equipment. For example, consider the effect of a 10 km duplex channel with 100 ms round trip latency. A fast PC running at 500 MHz clock rate would expend 50,000 clock cycles waiting for the attached device to respond, while a mainframe executing 1000 MIPS could expend over 100,000 instruction cycles in the same amount of time. The effect this may have on the end user depends on factors such as the application software. As another example, there is tremendous activity in burst mode routing and control traffic for next generation Internet (NGI) applications. In many cases, it is desirable for data to be transported from one point in the network to one or more other points in the least possible time. For some applications the time sensitivity is so important that minimum delay is the overriding factor for all protocol and equipment design decisions. A number of schemes have been proposed to meet this requirement,

including both signaling based and equipment intensive solutions employing network interface units (NIUs) and optical crossbar switches (OXBS).
In order to appreciate the impetus for designing burst mode switching
networks, it is useful to consider the delays that are encountered in wide
area data networks such as the Internet. For example, it takes 20.5 ms
for light to travel from San Francisco to New York City in a straight
line through an optical fiber, without considering any intervening equipment or a realistic route. Many transmission protocols require that data
packets traverse the network only after a circuit has been established; the
setup phase in TCP/IP, for example, involves three network traversals
(sending a SYN packet, responding with an ACK, and completing the
procedure with another ACK) before any data packets can be sent. While
this helps ensure reliable data transport, it also guarantees a minimum
network delay of more than 80 ms before data can be received. For some
applications this may be an unnecessary overhead to impose.

One way to mitigate the impact of network setup time on packet
latency is to pipeline the signaling messages. Two methods for accom-
plishing these time savings have been proposed [5, 6] which allow the
first packet delay to be reduced from two round trips to one and a half
round trips; these schemes are currently undergoing field trials. Another
paradigm for providing a burst mode switching capability using conventional equipment is to send a signaling packet prior to the actual
data packet transmission, and transmitting the data while the setup is in
progress. Some WDM systems have proposed using this approach [7],
using a dedicated wavelength for signal control information, a burst storage unit in order to buffer packets when necessary, and an implementation
of friendly link scheduling algorithm to insure efficient utilization of the
WDM channels. Another alternative is commonly known as "spoofing"
the channel; the attached channel extension equipment will be configured
to send ACKs prior to actually delivering the data. While this reduces
latency and improves performance, it also makes the assumption that
data can be delivered very reliably on the optical link; if there is a transmission problem after the attached channels have received their ACK of
successful delivery, the error recovery problem becomes quite difficult.
Although this approach has been implemented in commercial devices,
it has not been widely accepted because of the potential data integrity
exposures inherent in the design.

In many optical networks, cross-connect switches are employed which
act like electronically reconfigurable patch panels. The cross-connect
changes its switching state in response to external control information,

such as from an outband signal in the data network. The alternative to signaled data transmission is data switching using header information. This is somewhat different from a typical ESCON or gigabit Ethernet switch, which performs optical to electrical conversion, reads the data header, sets the switch accordingly, and then reconverts the data to the optical domain for transmission (sometimes retiming the signal to remove jitter in the process). An example of outband switching networks has been demonstrated [8]; in the approach, a data packet is sent simultaneously with a header that contains routing information. The headers are carried along with the data, but out of band, using subcarrier multiplexing or different wavelengths. At switch nodes inside the network, the optical signal is samples just prior to entering a short fiber delay line. While most of the signal is being delayed, a fast packet-processing engine determines the correct state of the switching fabric, based on the incoming header and a local forwarding table. The switch fabric is commanded to enter a net state just before the packet exists the delay line and enters the switch. This method provides the lower possible latency for packet transmission and removes the task of pre-calculating the arrival time of burst mode data packets.

6.4   Protection and Restoration

Backup fiber protection or restoration refers to the multiplexer's ability to support a secondary fiber path for redundancy in case of a fiber break or equipment failure. The most desirable is the so-called "1 + 1" SONET-type protection switching, the standard used by the telecom industry, in which the data is transmitted along both the primary and the backup paths simultaneously, and the data switches from the primary to the backup path within 50 ms (this is the SONET industry standard for voice communications; while it has been commonly adopted by protocol independent WDM devices, the effects of switching time on the attached equipment depends on the application, as in the previous discussion of latency effects). This is also known as an optical unidirectional protection switched ring (OUPSR); there are of course many variations, such as a bi-directional protection switched ring (BPSR). There are different meanings for protection, depending on whether the fiber itself or the fiber and electronics are redundant. In general, a fully protected system includes dual redundant cards and electrical paths for the data within the multiplexer, so that traffic is switched from the primary path to the secondary path not only if the fiber breaks, but also if a piece of equipment in the multiplexer fails. A less sophisticated but more cost effective option is the so-called "fiber trunk switch," which simply switches from the primary path to the backup path if the primary path breaks. Not all trunk switches monitor the backup path as $1 + 1$ switches do, so they can run the risk of switching traffic to a path which is not intact (some trunk switches provide the so-called "heartbeat" function, sending a light pulse down the backup path every second or so to establish that the backup link remains available). Trunk switches are also slower, typically taking from 100 ms to as much as 2 seconds to perform this switchover; this can be disruptive to data traffic. There are different types of switches; a unidirectional switch will only switch the broken fiber to its backup path (e.g. if the transmit fiber breaks, then the receive fiber will not switch at the same time). By contrast, a bi-directional switch will move both the transmit and the receive links if only one of the fibers is cut. Some types of datacom protocols can function properly only in a bi-directional switches environment because of timing dependencies on the attached equipment; the link may be required to maintain a constant delay for both the transmit and the received signal in a synchronous computer system, for example. Some switches will toggle between the primary and the secondary paths searching for a complete link, while others are non-revertive and will not return to the primary path once they have switched over. It is desirable to have the ability to switch a network on demand from the network management console, or to lock the data onto a single path and prevent switching (e.g. during link maintenance). Another desirable switching feature is the ability to protect individual channels on a per channel basis as the application requires; this is preferable to the "all or none" approach of trunk switches, which require that either all channels be protected, or no channels be protected. Hybrid schemes using both trunk switching and $1 + 1$ protection are generally not used, since they require some means of establishing a priority of which protection mechanism will switch first and since they defeat the purpose of the lower cost trunk switch. Other features such a dual redundant power supplies and cooling units with concurrent maintenance (the so-called "hot swappable" components, which can be replaced without powering down the device) should also be part of a high reliability installation.

Another desirable property is self-healing, which means that in the event of a fiber break of equipment failure the surviving network will continue to operate uninterrupted. This may be accomplished by re-routing traffic around the failed link elements; some form of protection switching or bypass switching can restore a network in this manner. In a larger network consisting of multiple cross-connects or add/drop multiplexers, there are two approaches to producing optical self-healing networks [9]. One is to configure a physical-mesh topology network with optical cross connects (OXCs), and the other is to configure a physical ring topology network with optical add/drop multiplexers (OADMs, or WDM ADMs). The OXCs used today are opaque (they require optical-to-electrical conversion before switching and electrical-to-optical conversion after switching); future designs call for transparent OXCs which would perform all optical switching. Various technologies for optical switching and OXCs have been suggested, including thermo-optic devices, liquid crystals, acousto-optic devices, inkjet or bubblejet technology, and micro-electromechanical (MEM) and micro-optoelectronic mechanical (MOEM) devices. Ring topology based optical self-healing networks are generally preferred as a first step because of the lower cost of OADMs compared with OXCs; also the protection speed in a ring topology is much faster, and the OADM is more transparent to data rate and format. Eventually, in future photonic networks, multiple optical self-healing rings may be constructed with emerging large scale OXCs. New architectures have also been proposed, such as bi-directional wavelength path-switched ring (BWPSR), which uses bi-directional wavelength-based protection and a wavelength-based protection trigger self-healing ring network [10].

6.5   Network Management

Some DWDM devices offer minimal network management capabilities, limited to a bank of colored lamps on the front panel; others offer sophisticated IP management and are configured similar to a router or switch. IP devices normally require attached PCs for setup (defining IP addresses, etc.) and maintenance; some can have a "dumb terminal" attachment to an IP device which simplifies the setup, but offers no backup or redundancy if the IP site fails. Some devices offer minimal information about the network; others offer an in-band or out-band service channel which carries management traffic for the entire network. When using IP management, it is important to consider the number of network gateway devices (typically routers or switches) which may be attached to the DWDM product and used to send management data and alarms to a remote location. More gateways are desirable for greater flexibility in managing the product. Also, router protocols such as Open Shortest Path First (OSPF) and Border Gateway Protocol (BGP) are desirable because they are more flexible than "static" routers; OSPF can be used to dynamically route information to multiple destinations. Many types of network management software are available, in datacom applications, these are often based on standard SNMP protocols supported by many applications such as HP Openview, CA Unicenter, and Tivoli Netview. IP network management can be somewhat complex, involving considerations such as the number of IP addresses, number of gateway elements, resolving address conflicts in an IP network, and other considerations. Telecom environments will often use the TL-1 standard command codes, either menu-driven or from a command line interface, and may require other management features to support a legacy network environment; this may include CLEI codes compatible with the TERKS system used by the telecommunications industry and administered by Telcordia Corp., formerly Bell Labs [11]. User-friendly network management is important for large DWDM networks, and facilitates network troubleshooting and installation; many systems require one or more personal computers to run network management applications; network management software may be provided with the DWDM, or may be required from another source. It is desirable to have the PC software pre-loaded to reduce time to installation (TTI), pre-commissioned and pre-provisioned with the users configuration data, and tested prior to shipment to insure it will arrive in working order. Finally, note that some protocol-specific WDM implementations also collect network traffic statistics, such as reporting the number of frames, SCSI read/write operations, MB payload per frame, loss of sync conditions, and code violations (data errors). Optical management of WDM may also include various forms of monitoring the physical layer, including average optical power per channel and power spectral density, in order to proactively detect near end-of-life components or optimize performance in amplified WDM networks.

References

[1] DeCusatis, C. (ed.) (2001) *Handbook of Fiber Optic Data Communication*, 2nd edition, Academic Press: New York.
[2] "Optical interface for multichannel systems with optical amplifiers", draft standard G.MCS, annex A4 of standard COM15-R-67-E, available from the International Telecommunication Union (1999).

[3] Ferries, M. (1999) "Recent developments in passive components and modules for future optical communication systems", p. 61, paper MI1, Proc. OSA annual meeting, Santa Clara, CA.

[4] See information on the Abilene network, http://www/abilene.internet2.edu.

[5] Meagher, B., Yang, X., Perreault, J. and MacFarland, R. (2000) "Burst mode optical data switching in WDM networks", paper FC3.2, Proc. IEEE summer topical meeting, Boca Raton, FL.

[6] Wei, J. (2000) "The role of DCN in Optical WDM Networks", Proc. OFC 2000, Baltimore, MD.

[7] Turnet, J.S. (2000) "WDM Burst switching for petabit data networks", Proc. OFC 2000, Baltimore, MD.

[8] Chang, G.K. *et al.* (2000) "A proof-of-concept, Ultra-low latency optical label switching testbed demonstration for next generation internet networks", Proc. OFC 2000, Baltimore, MD.

[9] Henmi, N. *et al.* (1998) "OADM workshop", EURESCOM P615, p. 36.

[10] Henmi, N. (2000) "Beyond terabit per second capacity optical core networks", paper FC2.5, Proc. IEEE summer topical meeting, Boca Raton, FL.

[11] Telcordia Technologies standard GR-485-CORE, "Common language equipment coding procedures and guidelines: generic requirements", issue 3, May 1999.

Further Reading

Brown, T. (2000) Optical Fibers and Fiber Optic Communications, in *Handbook of Optics*, vol. IV, Chapter 1, OSA Press.

Dutton, H. (1999) *Optical Communications*, Academic Press: San Diego.

Goff, D.R. (1999) *Fiber Optic Reference Guide*, 2nd edition, Focal Press: Boston, Mass.

Goralski, W.J. (2000) *SONET*, 2nd edition, McGraw-Hill: NY.

Kaminow, I.P. and Koch, T.L. (ed.) (1997) *Optical Fiber Telecommunications*, Academic Press: San Diego, CA.

Chapter 7 | Fiber Optic Communication Standards

There have been several international standards adopted for optical communications. This section presents a brief introduction to the concepts behind standardization, including some of the major standards bodies relevant to the fiber optic industry. We then provide an overview of several major optical communication standards, including the following:

- Enterprise System Connection/Serial Byte Connectivity (ESCON/SBCON)
- Fiber Distributed Data Interface (FDDI)
- Fibre Channel Standard and FICON (Fiber Connection)
- Asynchronous Transfer Mode (ATM)/Synchronous Optical Network (SONET)
- Gigabit and 10 Gigabit Ethernet
- InfiniBand.

7.1 Why Do We Need Standards?

One of the main goals of a standardization effort is to create a minimal set of requirements that will achieve a maximum amount of interoperability. To encourage innovation, it is important that the minimum standards not be overly restrictive; there are often multiple technical solutions to a design problem, and different design points may be desired for different applications. In this way, a well written standard offers more choices to the end user; the standard also insures that products from different vendors will interoperate to some degree, making it possible to trade off other factors such as cost. A standard is simply a documented agreement containing the technical specifications or other rules, guidelines, and definitions which will establish a common set of functions, if they are applied consistently. This differs from a product specification, which contains a much more detailed description and may exceed the standard limits in some areas, such as reliability or additional functionality. In fact, multiple products may claim compliance with a given standard, but may have very different implementations and specifications.

Traditional industry standards result from a consensus reached between multiple suppliers, users, and other groups such as government agencies; the majority of products or services available in the industry will conform to a successful standard. There is a large and often confusing array of formal industry standards groups and activities; we will provide only a high level overview of several major groups. Many standards are implemented by government regulations, including product safety and environmental protection standards. While we will not discuss these in detail, many optical communication products will be required to state formal compliance with areas such as fire safety, laser eye safety and the use of lead-free components in transceivers and cable assemblies (i.e. the Reduction of Hazardous Substances or RoHaS regulations, which stipulate the use of lead-free materials). These standards may require specific labeling on a product or its documentation (translated into representative languages), as well as maintaining updated records on file with government agencies responsible for enforcing the standards. This can create problems, however, when different countries or regions adopt different standards for similar products. For example, there are a number of standards governing fiber optic cable installation (see the building premise wiring standards from BICSI); in particular, the fire safety standards for optical cables vary in different parts of the world. In North America, plenum rated cables are required by the National Electrical Code and the National Fire Code for installation in air ducts; these cables are certified by independent test groups (such as Underwriters Labs, rating UL 910, OFNP) not to burn until extremely high temperatures are reached. Local fire marshals or insurance carriers may require this cable even in environments which are not specifically addressed in the standard, such as raised floors in computer rooms. In Europe and other countries, the standards allow cables to burn at a lower temperature than plenum materials, but is certified not to emit toxic fumes; these are known as low smoke, zero halogen cables. Other types of cables, such as riser rated (UL OFNR), offer lower fire protection but had previously been widely adopted as a de facto industry standard. The various standards make it necessary for optical cable manufacturers to offer a range of different products in different markets, maintain multiple part numbers for cables whose only difference is fire code rating, and contributes to confusion among users and installers. Note that additional standards on other aspects of fiber optic cables are available from the Electronic Industry Association/Telecom Industry Association (EIA/TIA) and from telecommunication industry associations such as BICSI [1, 2].

The existence of non-harmonized standards can form barriers to export and trade; hence, many international standards have been created by collaborative organizations. A leading example is the International Standards Organization (ISO), which has been involved in many standards activities across a wide range of industries, some of which apply to optical communications products or manufacturing processes (many industries are now familiar with the ISO 9000 family of quality standards, for example). The technical work of ISO is highly decentralized, carried out in a hierarchy of over 2700 technical committees and working groups. The responsibility for administering a standards committee is accepted by one of the national standards bodies that make up ISO; this includes organizations such as AFNOR, ANSI, CSBTS, and many more.

Another major international standards organization, the International Electrotechnical Commission (IEC), performs a similar role for its member organizations. Within the United States and North America, private sector voluntary standardization has relied heavily on the American National Standards Institute (ANSI) for over 80 years. This organization facilitates development of standards by establishing consensus among various groups; the standards themselves are created by technical experts, often representing professional societies such as the IEEE, Optical Society of America, SPIE, American Physical Society, etc. Through ANSI, these groups also have access to developing standards under discussion by ISO and IEC. There are still specific requirements for some geographic regions; notably, telecom equipment must comply with European standards governing radio frequency and electromagnetic interference, and must bear the Communaute Europeenne (CE) mark to verify they have been tested. Many countries still maintain their own standards for communications equipment; however, this requires vendors to obtain homologation and type approval for the import or export of certain fiber optic products. Despite the recent simplification created by the European

Community, these regulations remain a complex area, with the penalties for noncompliance including denial of import rights to certain nations.

There are other standards, not presented in detail here, which also play a significant role in optical link design. For example, the ITU is one of the oldest industry standards bodies [3] formed in 1865 under the name International Telegraph Union (about 21 years after Samuel Morse ushered in the telecommunications era by sending the first message over a public wire between Washington and Baltimore). Following the development of the telephone in 1876 and radio transmission in 1920, the organization went through a series of organizational changes in response to emerging fields of communication technology, including a merger with the International Telephone and Telegraph Consultative Committee (CCITT) in 1993. The modern ITU is organized into three main areas, namely telecommunication standardization (ITU-T), radio communication (ITU-R), and telecommunication development (ITU-D). Among their many contributions are the industry standard wavelength grids for wavelength division multiplexed optical communications and the generic frame procedure (GFP) for encapsulation of multiple data protocols under SONET [3].

In many cases, technical societies have taken a leading role in the establishment of new standards; for example, the IEEE is among the largest providers of communication standards as well as other documents which affect all aspects of the electrical engineering and computer science professions. Other groups, such as the more recently formed InfiniBand Trade Association (IBTA), have also been successful in using a similar approach. Various other standards related to fiber optics are available from organizations such as Bellcore, the National Research Council's Committee on Optical Science & Engineering, the Electronic Industries Association (EIA) and Electronic Industries Foundation (EIF, which publishes standardized Fiber Optic Test Procedures (FOTP) and the JEDEC series of manufacturing standards), the National Institute of Standards and Technology (NIST), and other groups [4]. While these efforts have enabled significant progress in the industry, they have been criticized recently as being too slow in their efforts to obtain a global consensus. Given the rapidly evolving pace of fiber optics technology, many vendors have chosen to collaborate informally in order to get products to market more quickly, only pursuing formal standardization afterwards. This has led to the formation of multi-source agreements (MSAs) through which a group of suppliers agrees to produce products according to documented guidelines. While MSAs are not enforced by any independent agency, companies who fail to comply will find themselves at a significant disadvantage in the marketplace, since many customers are demanding alternate sources for their optical components. Since they are formed more quickly than traditional standards, MSAs will not always succeed in anticipating market needs, and may dissolve over time; on the other hand, a broadly adopted MSA supported by large companies can effectively become a de facto standard. There are many MSAs in place today, including those for SFF fiber optic interfaces and packaging for high data rate optical transceivers [5]. These should not be confused with trade associations which are established simply to promote one possible design solution, and may not have broad based industry acceptance. For example, there are many different kinds of standardized optical connectors, each of which has a group of suppliers and users who advocate its use in a particular set of applications. Further confusing the distinction, some industry standards also specify the type of optical fiber and connectors suitable for a given application.

One of the unique requirements of fiber optic systems is eye safety for optical sources, including different types of lasers and LEDs used in communication systems. Most telecommunications equipment is maintained in a restricted environment and accessible only to personnel trained in the proper handling of high power optical sources. Datacom equipment is maintained in a computer center and must comply with international regulations for inherent eye safety; this limits the amount of optical power which can safely be launched into the fiber, and consequently limits the maximum distances which can be achieved without using repeaters or regenerators. For the same reason, datacom equipment must be rugged enough to withstand casual use while telecom equipment is more often handled by specially trained service personnel. Laser eye safety is defined by various industry standards [6], including the U.S. Food and Drug Administration Center for Devices and Radiological Health (FDA 21 CFR part 1040) which enforces the Radiation Control for Health and Safety Act of 1968 (Title 21, CFR subchapter J), the ANSI Z136.X standards, and the IEC (standard 60825-1, or European standard EN 60825-1). Certified laser safety training is available from various organizations, including the Laser Institute of America (LIA). These standards have a direct impact on the design of optical networks; since the maximum optical power launched into an optical fiber is determined by international laser eye safety standards, the number and separation between optical repeaters and regenerators is determined by the optical fiber link loss budgets.

For the remainder of this section, we will discuss the leading communication protocol standards used in fiber optics, as developed by ANSI, IEEE, and other bodies.

7.2   ESCON/SBCON

The ESCON* architecture was introduced on the IBM System/390 family of mainframe computers in 1990 as an alternative high speed I/O channel attachment [7, 8]. The ESCON interface specifications were adopted in 1996 by the ANSI X3T1 committee as the SBCON standard [9].

The ESCON/SBCON channel is a bi-directional, point-to-point 1300 nm fiber optic data link with a maximum data rate of 17 MB/s (200 Mbit/s). ESCON supports a maximum unrepeated distance of 3 km using 62.5 micron multimode fiber and LED transmitters with an 8 dB link budget, or a maximum unrepeated distance of 20 km using single-mode fiber and laser transmitters with a 14 dB link budget. The laser channels are also known as the ESCON Extended Distance Feature (XDF). Physical connection is provided by an ESCON duplex connector. Recently, the single-mode ESCON links have adopted the SC duplex connector as standardized by Fibre Channel. With the use of repeaters or switches, an ESCON link can be extended up to 3–5 times these distances; however, performance of the attached devices typically falls off quickly at longer distances due to the longer round-trip latency of the link, making this approach suitable only for applications which can tolerate a lower effective throughput, such as remote backup of data for disaster recovery. ESCON devices and CPUs may communicate directly through a channel-to-channel attachment, but more commonly attach to a central nonblocking dynamic crosspoint switch. The resulting network topology is similar to a star-wired ring, which provides both efficient bandwidth utilization and reduced cabling requirements. The switching function is provided by an ESCON Director, a nonblocking circuit switch. Although ESCON uses 8B/10B encoded data, it is not a packet switching network; instead, the data frame header includes a request for connection which is established by the Director for the duration of the data transfer. An ESCON data frame includes a header, payload of up to 1028 bytes of data, and a trailer. The header consists of a two character start-of-frame delimiter, 2 byte destination address, 2 byte source address, and one byte of link control information. The trailer is a two byte cyclic redundancy

*ESCON is a registered trademark of IBM Corporation, 1991.

check (CRC) for errors and a three character end-of-frame delimiter. ESCON uses a DC-balanced 8B/10B coding scheme developed by IBM.

7.3   Fiber Distributed Data Interface

The FDDI was among the first open networking standards to specify optical fiber. It was an outgrowth of the ANSI X3T9.5 committee proposal in 1982 for a high speed token passing ring as a back-end interface for storage devices. While interest in this application waned, FDDI found new applications as the backbone for local area networks (LANs). The FDDI standard was approved in 1992 as ISO standards IS 9314/1-2 and DIS 9314-3; it follows the architectural concepts of IEEE standard 802 (although it is controlled by ANSI, not IEEE, and therefore has a different numbering sequence) and is among the family of standards including token ring and Ethernet which are compatible with a common IEEE 802.2 interface. FDDI is a family of 4 specifications, namely the physical layer (PHY), physical media dependent (PMD), media access control (MAC), and station management (SMT). These four specifications correspond to sublayers of the Data Link and Physical Layer of the OSI reference model; as before, we will concentrate on the physical layer implementation.

The FDDI network is a 100 Mbit/s token passing ring, with dual counter-rotating rings for fault tolerance. The dual rings are independent fiber optic cables; the primary ring is used for data transmission, and the secondary ring is a backup in case a node or link on the primary ring fails. Bypass switches are also supported to re-route traffic around a damaged area of the network and prevent the ring from fragmenting in case of multiple node failures. The actual data rate is 125 Mbit/s, but this is reduced to an effective data rate of 100 Mbit/s by using a 4B/5B coding scheme. This high speed allows FDDI to be used as a backbone to encapsulate lower speed 4, 10, and 16 Mbit/s LAN protocols; existing Ethernet, token ring, or other LANs can be linked to an FDDI network via a bridge or router. Although FDDI data flows in a logical ring, a more typical physical layout is a star configuration with all nodes connected to a central hub or concentrator rather than to the backbone itself. There are two types of FDDI nodes, either dual attach (connected to both rings) or single attach; a network supports up to 500 dual attached nodes, 1000 single attached nodes, or an equivalent mix of the two types. FDDI specifies 1300 nm LED transmitters operating over 62.5 micron multimode fiber as the reference media, although the standard also provides for the attachment of 50, 100, 140, and 85 micron fiber. Using 62.5 micron fiber, a maximum distance of 2 km between nodes is supported with an 11 dB link budget; since each node acts like a repeater with its own phase lock loop to prevent jitter accumulation, the entire FDDI ring can be as large as 100 km. However, an FDDI link can fail due to either excessive attenuation or dispersion; for example, insertion of a bypass switch increases the link length and may cause dispersion errors even if the loss budget is within specifications. For most other applications, this does not occur because the dispersion penalty is included in the link budget calculations or the receiver sensitivity measurements. The physical interface is provided by a special Media Interface Connector (MIC). The connector has a set of three color-coded keys which are interchangable depending on the type of network connection; this is intended to prevent installation errors and assist in cable management.

An FDDI data frame varies in length, and contains up to 4500 eight-bit bytes, or octets, including a preamble, start of frame, frame control, destination address, data payload, CRC error check, and frame status/end of frame. Each node has a MAC sublayer that reviews all the data frames looking for its own destination address. When it finds a packet destined for its node, that frame is copied into local memory, a copy bit is turned on in the packet, and it is then sent on to the next node on the ring. When the packet returns to the station that originally sent it, the originator assumes that the packet was received if the copy bit is on; the originator will then delete the packet from the ring. As in the IEEE 802.5 token ring protocol, a special type of packet called a token circulates in one direction around the ring and a node can only transmit data when it holds the token. Each node observes a token retention time limit, and also keeps track of the elapsed time since it last received the token; nodes may be given the token in equal turns, or they can be given priority by receiving it more often or holding it longer after they receive it. This allows devices having different data requirements to be served appropriately.

Because of the flexibility built into the FDDI standard, many changes to the base standard have been proposed to allow interoperability with other standards, reduce costs, or extend FDDI into the MAN or WAN. These include a single-mode PMD layer for channel extensions up to 20–50 km. An alternative PMD provides for FDDI transmission over copper wire, either shielded or unshielded twisted pairs; this is known as copper distributed data interface, or CDDI. A new PMD is also being developed to adapt FDDI data packets for transfer over a SONET link by stuffing approximately 30 Mbit/s into each frame to make up for the data rate mismatch (we will discuss SONET as an ATM physical layer in a later section). An enhancement called FDDI-II uses time division multiplexing to divide the bandwidth between voice and data; it would accommodate isochronous, circuit-switched traffic as well as existing packet traffic. Recently, an option known as low cost (LC) FDDI has been adopted. This specification uses the more common SC duplex connector instead of the expensive MIC connectors, and a lower cost transceiver with a 9-pin footprint similar to the single-mode ESCON parts.

7.4   Fibre Channel Standard

Development of the ANSI Fibre Channel Standard (FC) began in 1988 under the X3T9.3 Working Group, as an outgrowth of the Intelligent Physical Protocol Enhanced Physical Project. The motivation for this work was to develop a scaleable standard for the attachment of both networking and I/O devices using the same drivers, ports, and adapters over a single channel at the highest speeds currently achievable. The standard applies to both copper and fiber optic media, and uses the English spelling "fibre" to denote both types of physical layers. In an effort to simplify equipment design, FC provides the means for a large number of existing upper level protocols (ULPs)—such as IP, SCI, and HIPPI—to operate over a variety of physical media. Different ULPs are mapped to FC constructs, encapsulated in FC frames, and transported across a network; this process remains transparent to the attached devices. The standard consists of five hierarchical layers [10], namely a physical layer, an encode/decode layer which has adopted the DC-balanced 8B/10B code, a framing protocol layer, a common services layer (at this time, no functions have been formally defined for this layer) and a protocol mapping layer to encapsulate ULPs into FC. The second layer defines the Fibre Channel data frame; frame size depends upon the implementation, and is variable up to 2148 bytes long. Each frame consists of a 4 byte start of frame delimiter, a 24 byte header, a 2112 byte payload containing from 0 to 64 bytes of optional headers and 0 to 2048 bytes of data, a 4 byte CRC and a 4 byte end-of-frame delimiter. In October 1994, the Fibre Channel physical and signaling interface standard FC-PH was approved as ANSI standard X3.230-1994.

Logically, Fibre Channel is a bi-directional point-to-point serial data link. Physically, there are many different media options (including copper, multimode fiber, and single-mode fiber) and three basic network topologies. The simplest, default topology is a point-to-point direct link between two devices, such as a CPU and a device controller. The second, Fibre Channel Arbitrated Loop (FC-AL), connects between 2 and 126 devices in a loop configuration. Hubs or switches are not required, and there is no dedicated loop controller; all nodes on the loop share the bandwidth and arbitrate for temporary control of the loop at any given time. Each node has equal opportunity to gain control of the loop and establish a communications path; once the node relinquishes control, a fairness algorithm insures that the same node cannot win control of the loop again until all other nodes have had a turn. As networks become larger, they may grow into the third topology, an interconnected switchable network or fabric in which all network management functions are taken over by a switching point, rather than each node. An analogy for a switched fabric is the telephone network; users specify an address (phone number) for a device with which they want to communicate, and the network provides them with an interconnection path. In theory, there is no limit to the number of nodes in a fabric; practically, there are only about 16 million unique addresses. Fibre Channel also defines three classes of connection service, which offer options such as guaranteed delivery of messages in the order they were sent and acknowledgment of received messages.

Fibre Channel provides for both single-mode and multimode fiber optic data links using longwave (1300 nm) lasers and LEDs as well as shortwave (780–850 nm) lasers. The physical connection is provided by an SC duplex connector, which is keyed to prevent misplugging of a multimode cable into a single-mode receptacle. This connector design has since been adopted by other standards, including ATM, low cost FDDI, and single-mode ESCON. The requirement for international class 1 laser safety is addressed using open fiber control (OFC) on some types of multimode links with shortwave lasers. This technique automatically senses when a full duplex link is interrupted, and turns off the laser transmitters on both ends to preserve laser safety. The lasers then transmits low duty cycle optical pulses until the link is reestablished; a handshake sequence then automatically reactivates the transmitters.

7.5 ATM/SONET

Developed by the ATM Forum, this protocol has promised to provide a common transport media for voice, data, video, and other types of multimedia. ATM is a high level protocol which can run over many different physical layers including copper; part of ATM's promise to merge voice and data traffic on a single network comes from plans to run ATM over the Synchronous Optical Network (SONET) transmission hierarchy developed for the telecommunications industry. SONET is really a family of standards defined by ANSI T1.105-1988 and T1.106-1988, as well as by several CCITT recommendations [11–14]. Several different data rates are defined as multiples of 51.84 Mbit/s, known as OC-1. The numerical part of the OC-level designation indicates a multiple of this fundamental data rate, thus 155 Mbit/s is called OC-3. The standard provides for seven incremental data rates, OC-3, OC-9, OC-12, OC-18, OC-24, OC-36, and OC-48 (2.48832 Gbit/s). Both single-mode links with laser sources and multimode links with LED sources are defined for OC-1 through OC-12; only single-mode laser links are defined for OC-18 and beyond. SONET also contains provisions to carry sub-OC-1 data rates, called virtual tributaries, which support telecom data rates including DS-1 (1.544 Mbit/s), DS-2 (6.312 Mbit/s), and 3.152 Mbit/s (DS1C). The basic SONET data frame is an array of 9 rows with 90 bytes per row, known as a synchronous-transport-signal level 1 (STS-1) frame. In an OC-1 system, an STS-1 frame is transmitted once every 125 microseconds (810 bytes per 125 microseconds yields 51.84 Mbit/s). The first 3 columns provide overhead functions such as identification, framing, error checking, and a pointer which identifies the start of the 87 byte data payload. The payload floats in the STS-1 frame, and may be split across two consecutive frames. Higher speeds can be obtained either by concatenation of N frames into an STS-Nc frame (the "c" stands for "concatenated") or by byte interleaved multiplexing of N frames into an STS-N frame.

ATM technology incorporates elements of both circuit and packet switching. All data is broken down into a 53 byte cell, which may be viewed as a short fixed length packet. Five bytes make up the header, providing a 48 byte payload. The header information contains routing information (cell addresses) in the form of virtual path and channel identifiers, a field to identify the payload type, an error check on the header information, and other flow control information. Cells are generated asynchronously; as the data source provides enough information to fill a cell, it is placed in the next available cell slot. There is no fixed relationship between the cells and a master clock, as in conventional time division multiplexing schemes; the flow of cells is driven by the bandwidth needs of the source. ATM provides bandwidth on demand; for example, in a client–server application the data may come in bursts; several data sources could share a common link by multiplexing during the idle intervals. Thus, the ATM adaptation layer (AAL) allows for both constant and variable bit rate services. The combination of transmission options is sometimes described as a pleiosynchronous network, meaning that it combines some features of multiplexing operations without requiring a fully synchronous implementation. Note that the fixed cell length allows the use of synchronous multiplexing and switching techniques, while the generation of cells on demand allows flexible use of the link bandwidth for different types of data, characteristic of packet switching. Higher level protocols may be required in an ATM network to insure that multiplexed cells arrive in the correct order, or to check the data payload for errors (given the typical high reliability and low BER of modern fiber optic technology, it was considered unnecessary overhead to replicate data error checks at each node of an ATM network). If an intermediate node in an ATM network detects an error in the cell header, cells may be discarded without notification to either end user. Although cell loss priority may be defined in the ATM header, for some applications the adoption of unacknowledged transmission may be a concern.

ATM data rates were intended to match SONET rates of 51, 155, and 622 Mbit/s; an FDDI compliant data rate of 100 Mbit/s was added, in order to facilitate emulation of different types of LAN traffic over ATM. In order to provide a low cost copper option and compatibility with 16 Mbit/s token ring LANs to the desktop, a 25 Mbit/s speed has also been approved. For premises wiring applications, ATM specifies the SC Duplex connector, color coded beige for multimode links and blue for single-mode links. At 155 Mbit/s, multimode ATM links support a maximum distance of 3 km while single-mode links support up to 20 km.

7.6   Gigabit Ethernet

Ethernet is an LAN communication standard originally developed for copper interconnections on a common data bus; it is an IEEE standard 802.3 [15]. The basic principle used in Ethernet is carrier sense multiple access with collision detection (CSMA/CD). Ethernet LANs are typically configured as a bus, often wired radially through a central hub. A device attached to the LAN that intends to transmit data must first sense whether another device is transmitting. If another device is already sending, then it must wait until the LAN is available; thus, the intention is that only one device will be using the LAN to send data at a given time. When one device is sending, all other attached devices receive the data and check to see if it is addressed to them; if it is not, then the data is discarded. If two devices attempt to send data at the same time (e.g. both devices may begin transmission at the same time after determining that the LAN is available; there is a gap between when one device starts to send and before another potential sender can detect that the LAN is in use), then a collision occurs. Using CSMA/CD as the MAC protocol, when a collision is detected attached devices will detect the collision and must wait for different lengths of time before attempting retransmission. Since it is not always certain that data will reach its destination without errors or that the sending device will know about lost data, each station on the LAN must operate an end-to-end protocol for error recovery and data integrity. Data frames begin with an 8 byte preamble used for determining start-of-frame and synchronization, and a header consisting of a 6 byte destination address, 6 byte source address, and 2 byte length field. User data may vary from 46 to 1500 bytes, with data shorter than the minimum length padded to fit the frame; the user data is followed by a 2 byte CRC error check. Thus, an Ethernet frame may range from 70 to 1524 bytes.

The original Ethernet standard, known also as 10Base-T (10 Mbit/s over unshielded twisted pair copper wires) was primarily a copper standard, although a specification using 850 nm LEDs was also available. Subsequent standardization efforts increased this data rate to 100 Mbit/s over the same copper media (100 Base-T), while once again offering an alternative fiber specification (100 Base-FX). Recently, the standard has continued to evolve with the development of gigabit Ethernet (1000 Base-FX), which will operate over fiber as the primary medium; this has the potential to be the first networking standard for which the implementation cost on fiber is lower than on copper media. Currently under development as IEEE 802.3z, the gigabit Ethernet standard is scheduled for final approval in late 1998. Gigabit Ethernet will include some changes to the MAC layer in addition to a completely new physical layer operating at 1.25 Gbit/s. Switches rather than hubs are expected to predominate, since at higher data rates throughput per end user and total network cost are both optimized by using switched rather than shared media. The minimum frame size has increased to 512 bytes; frames shorter than this are padded with idle characters (carrier extension). The maximum frame size remains unchanged, although devices may now transmit multiple frames in bursts rather than single frames for improved efficiency. The physical layer will use standard 8B/10B data encoding.

The standard does not specify a physical connector type for fiber; at this writing there are several proposals, including the SC duplex and various SFF connectors about the size of a standard RJ-45 jack.

Transceivers may be packaged as gigabit interface converters, or GBICs, which allows different optical or copper transceivers to be plugged onto the same host card. There is presently a concern with proposals to operate long wave (1300 nm) laser sources over both single-mode and multimode fiber. When a transmitter is optimized for a single-mode launch condition, it will underfill the multimode fiber; this causes some modes to be excited and propagate at different speeds than others, and the resulting differential mode delay significantly degrades link performance. One proposed solution involves the use of special optical cables with offset ferrules to simulate an equilibrium mode launch condition into multimode fiber, known as optical mode conditioners.

7.7   InfiniBand

The InfiniBand (IB) specification was developed to address the need for increasing bandwidth across multiple layers of interconnect technology, from backplanes to longer distance serial links, using both optical and copper media. This standard is controlled by the InfiniBand Trade Association (IBTA), an outgrowth of previous consortia including next generation I/O (NGIO), Scalable coherent interface (SCI), and HIPPI 6400 (now known as GSN). The IB physical layer is built upon a serial link, or lane, operating at 2.5 Gbit/s; multiple parallel lanes are combined to achieve higher bandwidths, including 4X (10 Gbit/s), 8X (20 Gbit/s), and 12X (30 Gbit/s). Higher speed interfaces are achieved by doubling the data rate of a serial lane to 5 Gbit/s; this in turn leads to double data rate (DDR) interfaces including 4X (20 Gbit/s), 8X (40 Gbit/s), and 12X (120 Gbit/s). A serial quad data rate (QDR) interface operating at 10 Gbit/s is also specified. The serial optical interfaces typically use standard duplex LC optical interfaces, while the parallel optical interfaces are based on variations of the MPO interface. Dual MPO connectors are used to provide a full duplex link. Optical links are specified for both multimode and single-mode fiber versions, except for the 12X interface which does not have a single-mode specification. IB links use 8B/10B encoding and link control characters. A comprehensive discussion of IB is beyond the scope of this section, and the interested reader is referred to the IBTA website or various references on this topic [16, 17].

References

[1] Electronics Industry Association/Telecommunications Industry Association (EIA/TIA) commercial building telecommunications cabling standard (EIA/TIA-568-A), Electronics Industry Association/Telecommunications Industry Association (EIA/TIA) detail specification for 62.5 micron core diameter/125 micron cladding diameter class 1a multimode graded index optical waveguide fibers (EIA/TIA-492AAAA), Electronics Industry Association/Telecommunications Industry Association (EIA/TIA) detail specification for class IV-a dispersion unshifted single-mode optical waveguide fibers used in communications systems (EIA/TIA-492BAAA), Electronics Industry Association, New York.

[2] BICSI telecommunications industry association & building premise wiring standards; www.bicsi.org.

[3] See the international telecommunications union (ITU) website, www.itu.int.

[4] See, for example, InfiniBand Trade Association website, www.infinibandta.org/home; NIST center for fiber optic research, Boulder, Colorado; www.boulder.nist.gov; Bellcore standards: www.bellcore.com.

[5] See, for example, www.xfp.org; www.xenpak.org; www.sff.org.

[6] United States laser safety standards are regulated by the Dept. of Health and Human Services (DHHS), Occupational Safety and Health Administration (OSHA), Food and Drug Administration (FDA), Code of Radiological Health (CDRH), 21 Code of Federal Regulations (CFR) subchapter J; the relevant standards are ANSI Z136.1, "Standard for the safe use of lasers" (1993 revision) and ANSI Z136.2, "Standard for the safe use of optical fiber communication systems utilizing laser diodes and LED sources" (1996–97 revision); elsewhere in the world, the relevant standard is International Electrotechnical Commission (IEC/CEI) 825 (1993 revision).

[7] Stigliani, D. (1997) "Enterprise systems connection fiber optic link", in *Handbook of Optoelectronics for Fiber Optic Data Communications*, C. DeCusatis, R. Lasky, D. Clement, and E. Mass (eds) Chapter 13, Academic Press, to be published.

[8] "ESCON I/O Interface Physical Layer Document" (IBM document number SA23-0394), IBM Corporation, Mechanicsburg, PA (third edition, 1995).

[9] ANSI Single Byte Command Code Sets CONnection architecture (SBCON), draft ANSI standard X3T11/95-469 (rev. 2.2) (1996).

[10] ANSI X3.230-1994 rev. 4.3, Fibre channel – physical and signaling interface (FC-PH), ANSI X3.272-199x, rev. 4.5, Fibre channel – arbitrated loop (FC-AL), June 1995, ANSI X3.269-199x, rev. 012, Fiber channel protocol for SCSI (FCP), May 30, 1995.

[11] ANSI T1.105-1988, Digital hierarchy optical rates and format specification.

[12] CCITT Recommendation G.707, Synchronous digital hierarchy bit rates.

[13] CCITT Recommendation G.708, Network node interfaces for the synchronous digital hierarchy.

[14] CCITT Recommendation G.709, Synchronous multiplexing structure.

[15] IEEE 802.3z, Draft supplement to carrier sense multiple access with collision detection (CSMA/CD) access method and physical layer specifications: media access control (MAC) parameters, physical layer, repeater and management parameters for 1000 Mb/s operation (June 1997).

[16] See, for example, www.infinibandta.org.

[17] DeCusatis, C. (ed.) (2001) *Handbook of Fiber Optic Data Communication*, Academic Press, New York (second edition).

Chapter 8 | Fabrication and Measurement

In this chapter we will discuss the standard equipment needed to fabricate optical fiber cables and components, including topics such as **fiber drawing, chemical vapor deposition (CVD), molecular beam epitaxy**, and **photolithography**.

There are many tests of fiber optics systems that can be done with standard laboratory equipment, such as the fiber optic test procedures referenced in the chapter on industry standards. For example, a **digital sampling oscilloscope (DSO)** used with a pattern generator can be used to characterize the waveform of an optical transmitter. However, in this section we will discuss several pieces of measurement equipment unique to fiber optics. These include the **fiber optic power meter** which measures how much light is coming out of the end of a fiber optic cable. **Optical time domain reflectometry (OTDR)** can non-destructively measure the location of signal loss in a link. BER can also be tested using a pattern generator and an error detector.

8.1 Fabrication Techniques

In this section, we will briefly describe the principles and apparatus used to fabricate optical fibers and related components. There have been many approaches over the years, some of which will be noted here for reference although they are less commonly used today.

8.1.1   FIBER DRAWING: LIQUID PHASE METHODS

Historically, fibers were drawn using crucible or liquid phase methods, also known as preform-free drawing. The double crucible method, as shown in Figure 8.1, produced fibers from glasses with low melting points by using a furnace with an outer crucible filled with cladding glass, and an inner crucible of core glass. Molten glass flows through two concentric

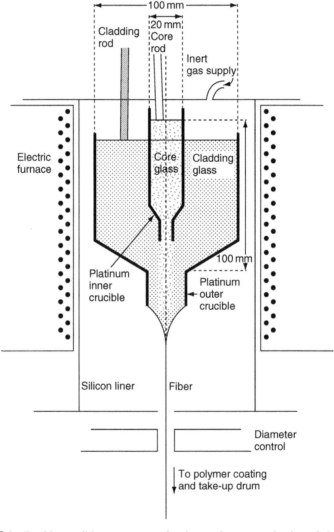

Figure 8.1   Double crucible arrangement for the continuous production of clad fibers.

nozzles in the base of each crucible, and forms a fiber core with a surrounding cladding; a filament is mechanically pulled from the melt to fabricate the clad fiber. Alternately, a seed crystal was inserted into the molten liquid in the crucible, and as it was slowly withdrawn a glass fiber would crystallize around it. Control of the core and cladding diameters was maintained by the fiber pulling speed and the head of molten glass in each crucible. By choosing suitable materials, index grading was possible by ionic diffusion. Another preform fabrication method, the "rod-in-tube" method, has the core and cladding melts cast separately and combined in a final melting/collapsing step. These techniques are not generally used in telecommunications grade silica fiber [1].

Preparation of optical fibers by glass melting methods is based on the use of powders which are premixed, heated in a crucible until they fuse, then agitated to produce a uniform mixture. The heat may be applied through black-body radiative coupling from the walls of an electrically heated furnace. Heat may be generated by coupling RF radiation to crucibles made of nonreactive, conductive materials, such as platinum, or coupled directly into the molten glass mixture (if it has the proper composition and is preheated to a high enough temperature that the mixture becomes conductive). The latter case helps prevent contamination of the crucible, since it remains relatively cold throughout the process.

8.1.2 FIBER DRAWING: CHEMICAL VAPOR DEPOSITION

Chemical vapor deposition is the primary technology used in modern fiber manufacturing. There are two common modern manufacturing processes, the "inside process," where a rotating silica tube is subjected to an internal flow of reactive gas and the "outside process," where a rotating, thin cylindrical target is used as the substrate for CVD, and must be removed before the bulk piece of glass, or boule, is sintered. For example, two gases, silicon chloride and oxygen, can react to produce silicon dioxide glass:

$$SiCl_4 + O_2 \rightarrow SiO_2 + 2Cl_2$$

These reactions occur at very high temperatures (1300–1600 °C), so it became necessary to rotate the tube containing the gas mixture to avoid having the tube deform. Other gases may also be present; for example, while helium is not specifically required for the deposition process, it is frequently used as a carrier gas for preform deposition, as a "sweep gas" to keep out impurities during preform consolidation, and as a heat transfer medium during fiber drawing.

There are two variants on the so-called "inside process." **Modified chemical vapor deposition (MCVD)** is basically a two-step process that involves fabrication of a specially constructed glass rod or **preform**, then melting the preform and drawing it into a fiber (Figure 8.2). The preform is a cylinder of silica about 10–20 cm in diameter and about 50–100 cm long. It consists of a core surrounded by a cladding with a desired refractive index profile, attenuation, and other characteristics, just like a larger version of the optical fiber. The gas mixture described previously is fed into the end of a rotating, heated silica tube. The chemical reaction forms glass particles, called **soot**, on the walls of the tube, which eventually build up to form a thin layer of glass. The process is repeated many times, and when the desired structure is formed the heat is increased to collapse the tube into a solid preform. In this way, the substrate is subjected to the internal flow of oxygen and $SiCl_4$ while an external torch provides energy for oxidation and the heat for sintering the deposited SiO_2. A final heating collapses the preform into an opaque **boule** so it is ready to be drawn into fibers. Collapse can take a significant fraction of the total processing time, and stability of the collapse plays an important role in determining the final geometry of the preform. The preforms fabricated as described above are pulled into glass fibers using a drawing tower, with a furnace at the top to melt the preform. The fiber diameter is automatically controlled to 125 microns with a 1 micron tolerance. During fiber drawing, an acrylate coating is applied to protect the pristine silica fiber from the environment.

Plasma-assisted chemical vapor deposition (PCVD) is an alternative process which provides the necessary energy to heat the glass by direct RF plasma excitation. Microwaves are used to form ionized gas plasma inside the tube; within this plasma, spots can acquire the equivalent of 60,000 cal energy, even though the mean temperature inside the furnace is only about 1200 °C. Thus, the electrons in the plasma move at very high speeds, and when they recombine considerable heat energy is released. This melts the soot particles obtained from the reaction between silica chloride, oxygen, and other catalyst gases; deposition of the desired glass occurs directly on a silica tube without the formation of soot. Thus, very thin layers can be formed since the process is not restricted by the size of the soot particles (about 0.1 micron diameter). This is a key advantage of the PCVD method. Next, there is a separate **sintering** step to provide a pore-free preform and remove impurities, and a final heating to collapse the preform into a state in which it is ready to be drawn.

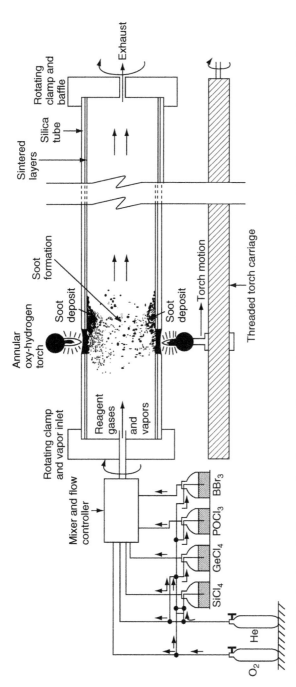

Figure 8.2 A schematic diagram of preform manufacture using internal chemical vapor deposition. The oxy-hydrogen torch is mounted on a carriage and moved along the tube at a speed which is servo controlled to maintain the desired temperature at the heated band. The gas and vapor flows are adjusted precisely for each pass of the torch.

There are also two variants on the "outside process." One is called vapor axial deposition (VAD), where the torch with chemical streams is stationary, and the rotating target is lifted as the deposition continues [2]. The second, called outside vapor deposition (OVD) uses a moving torch with chemical streams. The OVD method has three basic steps; laydown, consolidation, and draw. In the laydown step, a soot perform is deposited on the surface of a rotating target rod in a heated tube; the OVD method is known for this means of depositing the soot. The core material is deposited first, followed by the cladding material. Once the laydown process is complete, the target rod is removed from the preform (which is very porous), and the preform is then placed in a consolidation furnace. High temperatures are used to remove water vapor and sinter the preform into a solid boule. The hole left in the perform by the target rod is fused solid during the drawing step. The tip of the preform is heated to around 1900 °C until a piece of molten glass (sometimes called a gob) begins to fall from the preform. Gravity is allowed to draw the gob down into a thin strand of glass; as it is drawn, the glass is measured by laser micrometers up to 750 times per second, and controlled to precise tolerances. Two layers of protective coating are then applied, and the fiber is cured using ultraviolet light. A tractor belt winds the fiber onto a spool at speeds of 10–20 m/s. The completed fiber is then tested for tensile strength, transmission properties, and geometric tolerances. The OVD method has recently been used by such manufacturers as Corning and Lucent Technologies.

8.2   Fiber Bragg Gratings

Fiber Bragg Gratings were also discussed in previous chapters. When an optical fiber is exposed to ultraviolet light, the fiber's refractive index is changed; if the fiber is then heated or annealed for a few hours, the index changes can become permanent. The phenomena is called **photosensitivity** [3, 4]. In germanium-doped single-mode fibers, index differences between 10^{-3} and 10^{-5} have been obtained. Using this effect, periodic diffraction gratings can be written in the core of an optical fiber. This was first achieved by interference between light propagating along the fiber and its own reflection from the fiber endface [5]; this is known as the internal writing technique and the resulting gratings are known as **Hill gratings**. Another approach is the transverse holographic technique in which the fiber is irradiated from the side by two beams which intersect at an angle within the fiber core. Gratings can also be written in the fiber core by irradiating the fiber through a phase mask with a periodic structure. The fabrication of Fiber Bragg Gratings is discussed in greater detail in reference [4].

8.3   Semiconductor Device Fabrication

Light emitting diodes, lasers, and photodiodes have several manufacturing processes in common because they are all made from semiconductors, often grown in layers to form heterojunctions and quantum wells.

There are two major types of growth techniques. In **physical vapor deposition (PVD)**, physical means are used to deposit the films, such as bombarding the target with a plasma of ions (sputtering) or a molecular beam. In chemical techniques, the target is exposed to volatile chemicals, which react and deposit on the surface, as liquid phase epitaxy (LPE), CVD, and metal oxide chemical vapor deposition (MOCVD).

There are three major epitaxial growth techniques used to grow semiconductors: LPE, MOCVD, and **molecular beam epitaxy (MBE)**. Today, the vast majority of commercial semiconductors are grown by either MOCVD or MBE. Sputtering can also be used to deposit metal contacts using photolithography, but it is not used for layering the semiconductors.

Metal organic chemical vapor deposition (MOCVD) is the dominant technique in the growth of commercial semiconductor lasers because of its growth uniformity, high growth rate, and multiwafer growth capability. Typical growth sources used for MOCVD when growing III-V semiconductors (such as GaAs and InP) are trimethyl gallium (TMGa), trimethylaluminum (TMAl) and trimethylindium (TMIn) for the group III elements and arsine (AsH_3) or phosphine (PH_3) for the group V elements. These sources are transferred to the reactor by bubbling with hydrogen as a carrier gas. They are then heated in the reactor to temperatures from 500 to 800 °C, which decomposes them, and allows them to crystallize as III-V semiconductors on their substrate. The growth rate is driven by the amount of hydrogen carrying the V source. Dopants are available, and lattice matching is determined by adjusting the growth source ratio. In Figure 8.3 we see a schematic of a typical MOCVD system [6].

Molecular beam epitaxy offers a great deal of control because of its slow growth rate, coupled with in situ growth monitoring by reflection-high energy electron diffraction (RHEED). The molecular beam is made

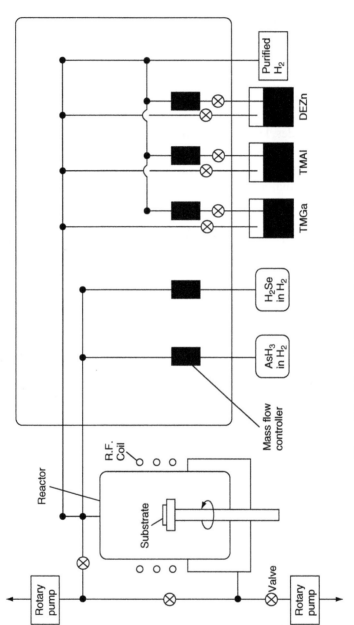

Figure 8.3 Schematic diagram of an MOCVD system.

by heating effusion cells filled with Ga, Al, As and doping sources, which are enclosed in liquid nitrogen cooled shrouds. The chemical composition of the film (e.g. the percentage of Al in AlGaAs) is controlled by varying the partial gas pressures of the sources (Ga, Al, As, etc.). The chamber is kept at an ultra-high vacuum (UHV) of 10^{-10} torr, and the substrate is rotated to achieve uniform thickness across the wafer. In Figure 8.4 we see a schematic of a MBE system. MBE has been used for the growth of commercial GaAs/AlGaAs laser structures, and has been demonstrated using other materials [6].

Photolithography is used to convert semiconductor wafers into devices. The first step is to apply photoresist to a clean surface. The photoresist can be positive, where application of UV makes the chemical structure of the resist soluble and easy to remove, or negative, where UV light makes the resist polymerized, and difficult to dissolve. The resist is most often applied by sputtering, creating a plasma that deposits on a wafer, which is spun to form an even coat. (It can also be applied by thermal evaporation.) Soft baking removes the solvents from the photo resist coat, and prepares it for receiving a mask.

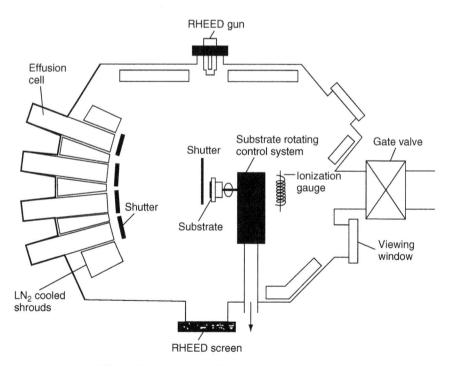

Figure 8.4   Schematic diagram of an MBE system.

A mask, or "photomask" is a square glass plate with a patterned emulsion of metal film on one side. This transfers the pattern to the photoresist. The mask can be placed in physical contact with the wafer, in close proximity to the wafer, or the mask can be avoided by using a projection printing technique, where the UV scanned across the surface is an image. Projection printers that step the mask image over the wafer surface are called step-and-repeat systems, and can attain 1 micron resolutions. The next step is development. The photoresist is dissolved where appropriate. Mesas are etched in the device. The final step is hard-baking. This hardens the photoresist and improves the adhesion of the photoresist to the wafer surface.

The above description of photolithography in some ways simplifies the multi-stage process a laser undergoes in production. In reference [6] a full description of the fabrication of a typical VCSEL is described, including multiple steps of photolithography – on the p-side photolithography is first used to define the initial mesa diameter, and later to define the VCSEL emission aperture on the mesa, while gold interconnect pads are also deposited by a photoresist liftoff process. The n-side also had Ni/AuGE/Au metal contacts deposited.

Several devices are usually formed on a single wafer. This wafer is either cleaved, which is split with a sharp instrument along the natural division in the crystal lattice, or sawed, to separate the individual chips. Edge-emitting lasers are often cleaved, VCSELs are often sawed [6].

8.4   Measurement Equipment

In this section, we describe some of the important optical measurement devices which may be used both in the laboratory and in commercial fiber installations to characterize an optical fiber link. These tools may be used when planning the installation of a new link, to determine whether the desired capacity can be achieved. They are also useful in fault isolation and troubleshooting of failures in optical links.

8.4.1 FIBER OPTIC POWER METER

The fiber optic power meter measures how much light is coming out of a fiber optic cable; it can be used to determine the amount of light being generated by an optical source, or the amount of light being coupled into an optical receiver. Optical power is usually measured in dBm, or decibels referenced to 1 mW. These devices measure the average optical power, not the peak power, so they are sensitive to the duty cycle of the data transmitted. Their wavelength and power range have to be appropriately matched to the system being measured. Most power meters used to test communication networks are designed to work at 850 nm, 1300 nm, and 1550 nm wavelength ranges and in the power range of -15 to -35 dBm for multimode links, or 0–40 dBM for single-mode links.

A test kit may include a fiber optic power meter and a test source. An optical loss test set (OLTS) includes a meter and source in one package, but since the source must be on one end of the cable, and the meter on the other, you need two of these in order to make a measurement.

A test source is a portable version of the transmitter attached to the fiber, and should match the wavelength and source type in the system. Generally, multimode fiber is tested with LEDs at 850 and 1300 nm, and single-mode fiber is tested with lasers at 1310 and 1550 nm.

Some cable measurements require a **reference cable**, also known as a test jumper. This is used to calibrate measurements to be made with an optical power meter, for comparison purposes with a fiber or component that may not be working properly. It is important to test these cables frequently, because their failure can invalidate an entire measurement procedure; it is also important to verify that the reference cable is matched properly to the size cable that is being tested (62.5/125, 50/125 or single mode).

8.4.2 OPTICAL TIME DOMAIN REFLECTOMETER (OTDR)

The OTDR looks at the backscatter signature of a fiber to determine loss. It can be used to find the location of damage to the fiber. It works like a radar of sonar, sending out a pulse from a very powerful laser. About one millionth of the light is back scattered from optical connectors, bends or fractures in the fiber, and similar discontinuities. This light is captured, averaged to improve the SNR, and analyzed.

Since the group velocity of light in the fiber is known, the round trip time for the OTDR pulse is expressed in distance. The OTDR displays backscatter vs. distance over the length of the fiber and any intermediate connections. For example, see the example in Figure 8.5. The slope of the trace is fiber attenuation, and a connector will show a peak, due to reflection, leading a lower value with the same slope. A splice, which usually is not reflective, will just be a dip leading to a lower value with the same slope.

There are several uncertainties and physical limitations in the OTDR measurement. The laser pulse width must be wide in order to induce

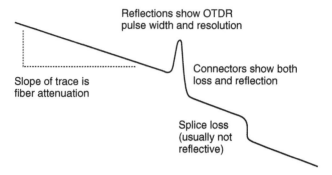

Figure 8.5   Information in the OTDR display.

enough backscattered light to process. This means the OTDR cannot distinguish most patchcords in a LAN cable plant or any events close to the OTDR itself. For example, when using an OTDR to nondestructively test a failed cable, it was determined that it broke near the connector. But since the connector was too close to the OTDR to be precise where the fracture was, the cable had to be destructively tested (opened up) to discover the flaw. Also, since the OTDR does not see the connectors on each end of the cables, it underestimates the actual cable plant loss.

Other uncertainties have a variety of causes. The variations in backscatter of fiber can cause major changes in the backscattered light, making splice or connector measurements uncertain by as much as ±0.4 dB. In multimode, cable the higher order modes are not filled by the OTDR laser, which makes it underestimate the loss of LED sources. Highly reflective events, such as connectors, can cause "ghosts" that are confusing.

However, the ability to non-destructively test fibers which are remote, buried, and otherwise inaccessible can be extremely valuable [7].

8.4.3   BIT ERROR RATE TEST (BERT)

While it is important that light reaches one end of the fiber to the other, a digital signal is composed of ones and zeros, and the fundamental measurement of the quality of a data communications system is found in the probability that the transmitted bits are correctly received. The Bit Error Ratio (BER) is the ratio of the number of bits received incorrectly compared to the number of bits transmitted.

In Figure 8.6 we see a BER test block diagram, where on one side of the device or system being tested is a pattern generator, and on the other side is an error detector. The pattern generator produces test patterns,

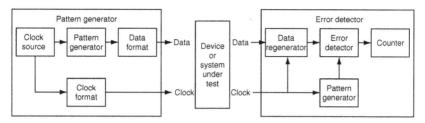

Figure 8.6 Bit error rate test block diagram.

some which simulate normal traffic, and some which intentionally stress some aspect of the system. The error detector determines whether the data received matches the pattern transmitted.

BER measurements primarily characterize system performance in terms of attenuation and signal power. However, any parameter that can impair system performance might be considered, such as jitter (Figure 8.7) [7].

8.4.4 DIGITAL SAMPLING OSCILLOSCOPE

As noted in the chapter on receivers, we can see that receiver sensitivity is specified at a given BER, which is often too low to measure directly in a reasonable amount of time (e.g. a 200 Mbit/s link operating at a BER

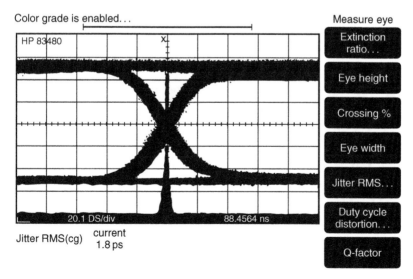

Figure 8.7 Using histograms to measure jitter.

of 10e-15 will only take one error every 57 days on average, and several hundred errors are recommended for a reasonable BER measurement). For practical reasons, the BER is typically measured at much higher error rates, where the data can be collected more quickly (such as 10e-4 to 10e-8) and then extrapolated to find the sensitivity at low BER.

We have made some assumptions about the receiver circuit. Most data links are asynchronous, and do not transmit a clock pulse along with the data; instead, a clock is extracted from the incoming data and used to retime the received data stream. We have made the assumption that the BER is measured with the clock at the center of the received data bit; ideally, this is when we compare the signal with a preset threshold to determine if a logical "1" or "0" was sent. When the clock is recovered from a receiver circuit such as a phase lock loop, there is always some uncertainty about the clock position; even if it is centered on the data bit, the relative clock position may drift over time. The region of the bit interval in the time domain where the BER is acceptable is called the eyewidth; if the clock timing is swept over the data bit using a delay generator, the BER will degrade near the edges of the eye window. Eyewidth measurements are an important parameter in link design, which can be measured with an DSO. This is a high speed oscilloscope used with a pattern generator to characterize transmitters. It shows you the output of the transmitter under different triggers.

A pattern trigger (see Figure 8.8) shows the sequence of ones and zeros. The relative separation of the logic levels, and rise and fall times can be quantified.

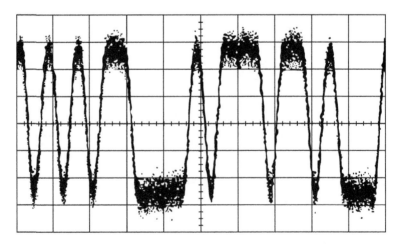

Figure 8.8   Pulse train from a high speed laser transmitter.

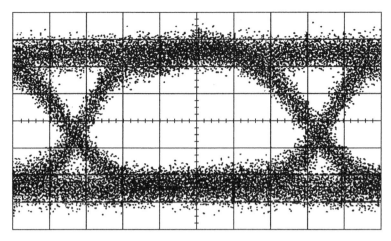

Figure 8.9 Eye-diagram displayed by a DSO.

Using the clock signal as the trigger input, the transmitted waveform can be sampled over virtually the entire data pattern generated by the transmitter. This forms an eye-diagram (see Figure 8.9). The eye diagram contains the following information about the transmitter: the relative separation between the logic levels, rise and fall times, jitter, eye height and width, and extinction ratio.

References

[1] Gower, J. (1984) *Optical Communications Methods*, pp. 90–113. Prentice Hall.

[2] Brown, T.G. (2000) "Optical Fibers and Fiber Optic Communications", in *Handbook of Optics*, Chapter 1, vol IV, pp. 1.44–1.47, OSA Press. McGraw-Hill.

[3] Hill, K. (2000) "Fiber Bragg Gratings", in *Handbook of Optics*, vol IV, Chapter 9, OSA Press. McGraw-Hill.

[4] Poulmellec, B., Niay, P., Douay, M., *et al.* (1996) "The UV induced refractive index grating in $Ge:SiO_2$ Preforms: Additional CW experiments and the macroscopic origins of the change in index". *J. Phys D. App. Phys*, 29:1842–1856.

[5] Hill, K.O., Malo, B., Bilodeau, F., *et al.* (1993) "Photosensitivity in optical fibers", *Ann. Rev. Mater. Sci.*, 23:125–157.

[6] Jiang, W. and Micahel S. Lebby (2002) "Semiconductor Laser and LED Fabrication;", in *Handbook of Fiber Optic Data Communication*, C. DeCusatis (ed.), Chapter 16, pp. 603–644. Academic Press.

[7] Hayes, J. and Greg, L. (2002) Testing Fiber Optic Local Area Networks *Handbook of Fiber Optic Data Communication*, C. DeCusatis (ed.), Chapter 9, pp. 333–363. Academic Press.

[8] For more information on fiber manufacturing, see:
- http://www.corningcablesystems.com/web/college/fibertutorial.nsf.
- http://www.thefoa.org/tech/fibr-mfg.htm.

Chapter 9 | Medical Applications

The use of optical light guides for viewing the human body dates back to the late 1880s, when Dr. Roth and Prof. Reuss of Vienna used a bent glass rod to illuminate body cavities, and David Smith of Indianapolis patented a dental illuminator using glass rods (for more on the history of medial optics, see [1]). In more modern times, non-invasive imaging by means of sonograms, magnetic resonance imaging, and improved x-rays have revolutionized medicine. **Endoscopy** offers the additional ability to reach inside the human body in a minimally invasive way. There are two major types of endoscopes: rigid endoscopes, which are sometimes called laparoscopes and are basically medical periscopes, and flexible endoscopes, which are either fiber-optic or CCD based. However, all endoscopes have benefited from fiber optic technology, because they use optical fibers to illuminate their image. Some devices also use fibers to perform remote laser surgery.

In the following section we will describe fiber optic endoscopes, laser fibers, medical illuminators, and biosensors. Our focus in this chapter will be on medical applications, although all these products have broader application. For example, while fiber optic endoscopes and laser fibers do have industrial applications (e.g. inspection of pipes or duct work which is difficult to reach in other ways), our emphasis will be on medical applications. Similarly, fiber optic lighting systems have many commercial applications; they are used for emergency lighting, automotive lighting, traffic signals, signage, lighting sensors, commercial, and decorative lighting. They are also very useful in medical applications, coupled with fiberscopes or microscopes. Fiber optic lighting systems can also be placed in high magnetic fields, such as near MRI machines, and can be used in stand-alone applications in this manner. While fiber sensors have broader industrial applications, especially in hard-to-reach places and hazardous areas where there is a danger from electrical sparks igniting explosions, they also have medical applications, especially working in conjunction with endoscopes.

9.1 Endoscopes

The use of fiber optics in endoscopy is perhaps the most significant application of fiber optics to the field of medicine. From encephaloscopes, which are used for examining cavities in the brain, to arthroscopes, which are used for diagnosing and treating knee problems, the human body can be directly probed with these devices. This approach has become part of the mainstream medical community; for example, all people 50 years or older are recommended to be screened for colorectal cancer, by either flexible sigmoidoscopy every five years or colonoscopy every ten years, and increased risk patients are recommended for screening younger and more frequently [2].

There are two major types of endoscopes: flexible ones, that use a fiber bundle or a CCD chip detector, and rigid laparoscopes, which are more like a rigid periscope. Flexible scopes are used in applications like colonoscopy where the probe is threaded through a winding path. Often the whole path is a part of the examination. Rigid scopes tend to be used in gynecology and arthroscopy, where a small incision allows the scope to be placed at the target where the procedure is performed. An advantage of laparoscopes is that they offer greater resolution and the absence of a visible lattice structure in the image (which is an artifact from the fiber bundles.) Nearly all fiber optic endoscopes include two different types of optical fibers: one used for the image and the other as a light guide to provide illumination. Additionally, other channels may be present for the passage of air, water, and remote control implements such as biopsy forceps and cytology brushes, as well as laser fibers.

There are two distinct categories of fiber based endoscopes: standard-sized devices such as gastroscopes, colonscopes, and bronchoscopes, and ultrathin ones used in needlescopes, ophthalmic endoscopes, and angioscopes. The ultrathin needlescopes contain 2000–6000 pixels with a diameter of about 0.2–0.5 mm [3]. The development of lenses for the copier industry has resulted in compact gradient-index lenses which are also used in small diameter endoscopes (<2 mm O.D.) [4]. For some applications, such as flexible sigmoidoscopy, there is a choice between a fiberoptic system and a video pathway system. The diameter of the insertion tube is greater with the video system, and it is twice as expensive, but it provides clearer images and allows a video record to be obtained [5].

Taking a rather simplified view, a fiberoptic endoscope contains three basic parts: an objective lens system that is inserted in the body and produces the image, a two-dimensional array of clad fibers that carries the image (called a **fiber relay**), and an ocular lens where the image is viewed. Illumination fibers are almost always necessary, since there is no other light inside the body (see Figure 9.1). Of course, actual design of an endoscope is quite a bit more complex—the relay lens, between the objective lens and the fiber relay, significantly affects image propagation [4]. The difference between a normal fiber bundle and an imaging bundle is that in the case of a bundle used for imaging, individual fibers have the same geometrical arrangement on each end of the bundle. This is called a **coherent fiber bundle** or **image bundle**. This can be done by either fusing only each end of the bundle, which is more flexible, or fusing the bundles over the entire length, which restricts their motion [6].

For example, a flexible sigmoidoscope (used for colon examination) consists of a control head, an insertion tube, and an umbilical cord. The control head houses the eyepiece, deflection knobs, valves (air-water and suction), and biopsy channel. An adaptor over the eyepiece allows you to view the image on a television monitor. The lens can be washed with water during the examination, and suction can be used to remove mucus and other contamination which may obscure the image [5].

Some important factors in the optical design of an endoscope include image position (focal length), brightness, and field of view. The length of the relay lens is adjusted for the desired endoscope length. The brightness of the image is determined by the **optical invariant** of the relay section, which is proportional to the numerical aperture and diameter of the relay lens. Current lenses limit the field of view to about 55°. This may be

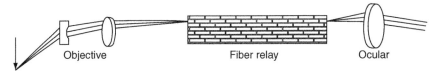

Objective   Fiber relay   Ocular

Figure 9.1 Optical layout of a fiberoptic endoscope.

increased by adding a negative lens to the tip of the endoscope, but there are drawbacks to that approach, such as smaller images and off-parallel imaging. Endoscope prisms may also be used for inclining the field of view (i.e. viewing to the side instead of straight ahead) [4].

The optical fibers used in flexible endoscopes typically consist of **fiber bundles**, also called **leached image bundles**. These are fused multi-fiber rods composed of a large number of glass fibers with their own individual claddings. The fiber cores can be as small as 8 microns in diameter, bundled on 10 micron center-to-center spacing. Each fiber represents one pixel in the resulting image; for example, a 5 × 5 fiber rectangular array bundled into a 50 micron outer jacket would yield a 25 pixel image, and multiple fiber bundles with perhaps several thousand elements would be used for higher resolution. The primary fiber cladding is an acid-resistant glass of relatively low refractive index. There is also a secondary cladding, an acid soluble glass which acts as a bonding agent for the individually clad fiber elements during the early stages of the manufacturing process. Image bundles are produced from one or more fiber drawing processes. The fibers are wound onto a drum one layer at a time, and assembled using a separate laminating operation. Multi-fiber rods are then cut to the desired length, and the endfaces are optically polished. The image bundle is then subjected to an acid bath, which leaches away the secondary cladding. This process removes the bonding between individual fibers along most of their length, producing a high degree of flexibility between the two ends of the bundle. The bundle may be used to record still images or video, and may contain extra channels for illumination or operating instruments. The bundle may have a numerical aperture of about 0.6 mm and typical lengths are around 600 mm (though in some cases, bundles as long as 4500 mm have been used). Several different types of leached image bundles are described in Table 9.1. Modifications of the basic

Table 9.1 **Typical leached image bundles of fiber**

Bundle outer diameter (mm)	Maximum length (mm)	Fiber core size (microns)	Fiber count
0.67	1000	8.9	About 10,000
1.10	1000	7.8	About 18,000
1.20	1000	8.4	About 18,000
1.50	2000	10.6	About 18,000
1.65	2350	11.6	About 18,000

design are used to magnify, minify, or otherwise enhance image quality. For example, tapering the fiber ends can provide magnification ratios up to 3:1, with a numerical aperture up to 1.0; the large end of the taper can be from 6 to 25 microns in diameter, while the small end can be from 3 to 6 microns in diameter. Different fiber endfaces can be machined (round, square, or rectangular) and different types are designed for use in the visible or near infra-red spectrum.

9.1.1   RIGID ENDOSCOPES

In the early days of invasive imaging (1950s), gastroenterologists were allowed to perform colonoscopies but not esophagoscopies, which could only be performed by specialized technicians according to the medical rules of that time. To allow gastroenterologists to use these devices, the Wolf-Schindler gastroscope was designed with optics which were fixed at a right angle to the long axis of the instrument, such that it was not possible to see the esophagus as the scope passed through the stomach. This design change allowed gastroenterologists to use the device for the first time [7].

The Wolf-Schindler gastroscope is an example of a rigid endoscope. The pioneer of rigid endoscopes was Rudolf Schindler, who first introduced a semiflexible gastroscope in 1932 [8], far before the first fiberoptic endoscope was developed by Hirschowitz in 1957 [9]. However, fiber optic technology also made rigid endoscopes far more practical by replacing the incandescent bulb at the distal tip with an illuminating fiber.

The rigid endoscope uses a series of relay lenses to pass the information in a compact way between the object on the distal end of the scope and the eyepiece (Figure 9.2). These relay lenses are rod lenses of the Hopkins design [7] with a length to diameter ratio as high as 10. The objective lens determines the field of view. The goal is to have as large a field of view as possible, while still working well with the relay lenses. The large

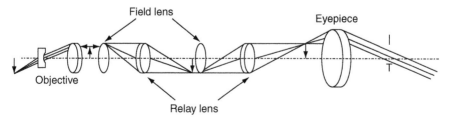

Figure 9.2 Optical layout of a rigid endoscope.

field angles in modern endoscopes lead to a small entrance pupil, which allows straightforward coupling of the entire ray bundle.

9.1.2   THE CLINICAL ENVIRONMENT

The clinical environment poses special challenges to fiber optic equipment design that are very different from a communications environment. For example, mechanical fragility poses a problem. When the needle at the end of an endoscope is bent, stress at the cement joints can separate the lenses. Since gradient-index endoscopes are very thin and have long glass lengths, they are particularly susceptible to breakage and lens separation.

However, the most significant problem is sterilization (the killing of bacteria and spores on the endoscope). Most surgical instruments are sterilized. The standard technique involves a combination of high temperature and pressure by means of an autoclave before each use. It would be desirable if an endoscope could be autoclaved at least 100 times during its useful lifetime. However, normal adhesives separate at autoclave temperatures, and even high temperature adhesives do not protect from steam entering and fogging the optics. Thus, standard autoclave methods do not work well for endoscopes. Currently the only clinical method that truly sterilizes endoscopes is exposing them to the poisonous environment of ethylene oxide. Unfortunately, the process is slow; the endoscope must be isolated until the gas dissipates, which takes over a day. More commonly, the endoscope is disinfected by soaking it in a solution of activated glutaraldehyde. This technique does not kill spores and is not permitted in all hospitals [4]. Some endoscopes, such as flexible sigmoidoscopes, may be used with a disposable sheath that protects the working surfaces from contamination [5]. This is not always possible, so the search for a truly reusable scope is ongoing.

9.2   Laser Fibers

Fibers are used as a remote delivery system for surgical lasers. It is also possible to dope an optical fiber with low levels of rare earth elements, so that it amplifies light or acts as a laser. The term **laser fiber** is used loosely in the technical literature when referring to both of these designs, although only the second implementation involves making a laser out of a length of optical fiber. When laser fibers are connected to endoscopes, they allow doctors to perform remote surgery in places difficult to reach by conventional means. This technique can reduce postoperative pain and recovery times significantly, because less tissue is damaged reaching the surgical site. There are many different types of lasers and wavelengths used for different medical procedures.

The typical of lasers used in laser fibers are very different from those used in communications or other applications of fiber optics. They include Holmium:yttrium-aluminum-garnet (Ho:YAG) laser, Niodinium: yttrium-aluminum-garnet (Nd:YAG), KTP (potassium titanyl phosphate), Argon lasers and some high power diode lasers. Excimer and CO_2 lasers are also used for laser surgery, but they are not coupled with fibers.

Ho:Yag lasers are considered the most versatile surgical lasers currently available. Their 2100 nm wavelength (mid infrared) is preferentially absorbed by water, and is typically absorbed within 0.5 mm of the tissue surface. This relatively low depth of thermal penetration and convenient delivery system that allows access in even the tightest places makes this laser suitable for a variety of surgeries. The Ho:YAG can vaporize, ablate, and coagulate both soft tissues and extremely hard materials. It is used for orthopedic, urologic, ENT, gynecologic, gastroenterologic, and general surgery. The laser cavity is a YAG crystalline rod doped with Ho, and possibly also thulium (Tm) and chromium (Cr), the later two improving the laser's efficiency. Energy is emitted in a train of pulses; both the energy per pulse and the pulse rate may be adjusted. Energy leaving the laser tube through a partially reflecting mirror is directed into a flexible optical fiber. There may also be special tips, to focus the beam into a small spot at a known working distance, and a micromanipulator. Nd:YAG lasers have a 1064 nm wavelength (near infrared) and are typically scattered over a 5 mm depth of tissue. They are used to cause photocoagulation or thermal denaturation of the tissue. Typical applications include urology, angioplasty, bronchopulmonary, gastroenterogy, and neurosurgery. Nd:YAG systems are similar to Ho:YAG systems, except that the laser cavity is an YAG crystalline rod doped with neodymium.

Since infrared and near infrared lasers are not visible, they are aimed using a second, visible helium–neon (He–Ne) laser, a laser diode, or a xenon lamp, which emits visible light (typically red) and follows the same path as the infrared laser beam. Some lasers operate in the visible spectrum; for example, KTP lasers have a wavelength of 532 nm (green). A Nd:YAG laser transmits its output through a KTP crystal, which converts the 1064 nm energy to half its wavelength. It can be designed so that it switches between 1064 and 532 nm, depending on the application.

Argon lasers have a wavelength of between 488 and 514 nm, which is blue/green visible light. Argon lasers have a laser tube filled with an argon gas mixture, which emits light when stimulated by an electric field. The energy leaves the laser tube through a partially reflecting mirror and is directed into a flexible optical fiber. Argon lasers can emit a single pulse or a train of pulses. Green light is preferentially absorbed by pigmented tissue and hemoglobin, and is typically absorbed within 3 mm of the tissue surface. Typical applications include bronchopulmonary, gastroenterology, dermatology, urology and high-power ophthalmic. Diode lasers use an array of solid state diode lasers emitting at near infrared light. Diode laser physics has been discussed in earlier sections on communications. They produce 60 W or more of power, and are used in ophthalmic and dermatological applications [10]. Note that optical fibers used for medical applications must be capable of carrying different wavelengths, including large parts of the visible spectrum, than those used for communication systems.

9.2.1 APPLICATIONS

Fiber lasers can be used to correct imperfections in the focal properties of the human eye. These include refractive errors such as myopia (near sightedness), hypermyopia (far sightedness), and astigmatism. One example is the increasingly popular surgical technique known as light assisted in situ keratamileusis (LASIK), which used lasers to permanently reshape the cornea [11]. Since the cornea adjusts the shape of the eye's lens, in much the same way that the lens of a camera is focused, removal of corneal tissue can change the eye's focal properties. Fiber lasers are used in other types of eye surgery as well. For example, diabetic retinopathy is a severe disorder of the retina that causes blood vessels to leak; this can be treated using a blue/green argon laser, whose wavelengths are preferentially absorbed by cells lying under the retina and by red hemoglobin in blood. Fiber lasers can also be used to treat glaucoma, a leading cause of blindness, by creating tiny openings in the iris to allow excess fluids inside the eye to drain and relieve pressure. A similar treatment is used for macular degeneration, which affects central vision in older adults; the laser light can destroy abnormal blood vessels so that scarring will not damage the patient's vision.

Fiber lasers have also found extensive applications in dental surgery [12]. Small amounts of light are directed through a fiber which can be placed between the teeth and gums. The laser energy can be directed to remove only diseased tissue and aids in reducing the bacteria associated with gum disease. There is less recurrence of gum disease following these treatments, compared with conventional dental surgery. The laser also loosens tartar buildup, which is associated with inflammation and bleeding gums. Higher powered fiber lasers are used for cutting and shaping hard tissues inside the mouth; for example, adjusting the gingival proportions to match the upper lip. Diode lasers can be used to reshape soft tissues in the mouth, as well. Laser cutting of tissue minimizes the risk of infection and avoids bleeding in the mouth during surgery; it can also be used to seal the gap between teeth and gums, forming a natural band-aid. Increasingly, cavities are being filled with materials that can be cured or hardened after exposure to laser light, particularly for the front teeth. This is administered from a hand-held tool using optical fibers to guide a 450 nm argon laser beam; fiber rods or tapers with high transmission at this wavelength are required. The transmission characteristics of this glass can remain quite stable, even after several thousand autoclave sterilization cycles.

Surgical lasers with fiber delivery systems use light with high intensity and narrow spot size, for example to treat various stages of cancer. Laser therapy is often given through a flexible endoscope inserted into an opening in the body, such as the mouth, nose, or a conventional surgical incision. This method is most often used to treat superficial cancers (on the surface of the body or the lining of internal organs), such as basal cell skin cancer and the early stages of cervical or lung cancer. Precisely delivered laser light can shrink or destroy tumors, and can also relieve symptoms associated with cancer such as bleeding or obstructions. For example, fiber lasers can shrink tumors blocking the windpipe or throat, and remove polyps blocking the colon or stomach.

Endoscope treatments typically use either carbon dioxide (CO_2) lasers, argon lasers, or Nd:YAG lasers. In particular, CO_2 and argon lasers can cut the skin surface without penetrating into deeper layers, while Nd:YAG lasers are more commonly applied to internal organs. While surgical lasers are used to cut and vaporize tissue, there is a significant dermatological and aesthetic application, too. In those cases, selective photolysis allows the laser to destroy dark structures, such as hair and pigment (tattoos), through selective heating while preserving the adjacent tissue.

Laser-induced interstitial thermotherapy (LITT), also called interstitial laser photocoagulation, involves inserting an optical fiber into a tumor, where laser light emitted from the fiber tip can damage or destroy abnormal cells. This is used to treat some cancers or to shrink tumors in the liver. The Nd:YAG lasers are most typically used through optical fibers for this treatment. Another type of cancer treatment is photodynamic therapy (PDT). In this case, a drug called photosensitizing agent is injected into the patient and absorbed by cells throughout the body; after a few days, the agent is found mostly in cancer cells. Laser light is then delivered to activate the agent and destroy the cancer cells. Because the photosensitizing agent makes the skin and eyes sensitive to light for about six weeks, patients are advised to avoid direct sunlight or bright indoor lights during this time. Argon lasers are often used to activate the drugs used in PDT.

In additional to the human applications, they are also used in veterinary medicine. For example, the Clark Equine Clinic advertises "Many surgical procedures using an endoscope guided laser are available such as to repair respiratory problems and to remove endometrial cysts in mares" [13].

Urologists report that treating an enlarged prostate by tradition means, transurethral resection of the prostate (TURP), requires a one-to-four-day hospital stay followed by four to five weeks of recovery, including wearing a catheter for one to three days. By comparison, the laser alternative, GreenLight PVP™ Laser Procedure for Photoselective Vaporization of the Prostate, typically requires no overnight stay and many patients go home a short time after the procedure [14, 15].

9.3   Illumination

Fiber optic illuminators for medical and industrial applications often use fiber bundles. The input cross section is arranged to match the light source, and the output can be arranged to meet the desired illumination needs. Unlike image bundles, maintaining coherence is unnecessary. The desire for an even distribution of light leads to "**fiber randomizing**," or distributing light fibers throughout the bundle, rather than in one part of it (Figure 9.3).

However, when used in image bundles, or with sensor fibers which have both sending fibers and detector fibers, the illumination fibers may be segregated in one part of the bundle. In Figure 9.4 we can see some differing geometries. While plastic fibers can be used for illumination, when combined with detector or sensor fibers, glass and quartz fibers must be used. This is due to lower transmission in certain spectral regions (higher UV and IR absorption), induced fluorescence in the fiber, and broadband interference. The fibers are coated with an electrolytic agent

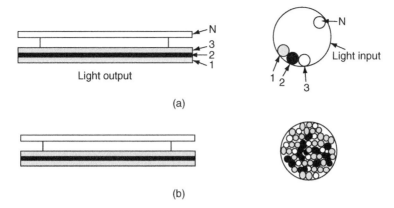

Figure 9.3 Possible fiber distributions or fiber randomizing for a line converter (sketch) (a) coarse randomization (b) fine randomization.

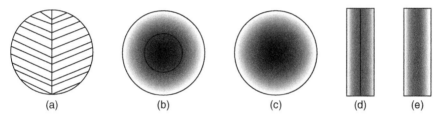

Figure 9.4 (a), (b), and (d) Segregated fiber cross sections. (c) and (e) Sender/Reciever randomized.

to prevent static buildup, and additional buffer and jacket coatings. Some of these coatings will affect the temperature limits of the device, which is especially important for medical applications which may require repeated sterilization cycles [6].

9.4 Biosensors

Fiber optics sensors and fiber Bragg gratings have been mentioned in previous chapters. Fiber Bragg Gratings are spectral filters fabricated within segments of optical fibers. Since a change in external conditions, such as temperature or mechanical force, will expand or compress the grating, changing its reflected wavelength, these sensors can be used in applications which measure changes in temperature and pressure. In addition to biological applications, this can be used for industrial applications such as measuring oil wells.

A biosensor is a probe that integrates a biological component, such as a whole bacterium or a biological product (e.g. an enzyme or antibody) with an electronic component to yield a measurable signal. An artificial electronic nose would be an example of a biosensor. Fiber optic sensors can be used as biosensors, and fiber optics can be integrated into other types of biosensors.

Optical fibers are used in biosensors for the measurement and monitoring of temperature, blood pressure, blood flow, and oxygen saturation levels [1]. Food is tested for pathogens using fiber optic probes which utilize fluorescent dyes [16]. They are also used for farm, garden, and veterinary analysis, fermentation process control, and various pharmaceutical applications. The advantage of using a biological component as a sensing element is their remarkable ability to distinguish between the substance we wish to analyze and similar substances. This makes it possible to measure specific chemicals with great accuracy. Biosensors also typically yield results very quickly, and can easily be integrated into a compact package. For example, special optical fibers with 500 micron diameter can be bundled together, and the tip of each fiber can be coated with an active biosensor, such as a reagent for glucose measurements. The glucose level in a blood sample can then be measured simply by dipping the fiber tip into the sample and illuminating the fiber tips with light, which will change color in response to the glucose concentration.

A specific bio-sensor technique is laser fluorescence diagnosis, which was developed by Tuan Vo-Dinh in collaboration with medical researchers at Thompson Cancer Survival Center in Knoxville. When the esophagus interior is illuminated with 410 nm blue light, normal tissue cells fluoresce at different frequencies than cancer cells. This device, which is swallowed by the patient, eliminates the need for painful biopsies and boasts a 98% accuracy (from tests on 200 patients), compared to biopsies. Vo-Dinh's group is also studying ways to apply optical techniques to detect skin, cervical, and colon cancers, as well as monitoring the status of diabetes by illuminating the eyeball [17].

The basic elements of a biosensor system include a light source, light delivery means, object for light to interact with, and light acquisition and detection systems (see Figure 9.5). There are three possible configurations: transmission, fluorescence, and reflection. While our mental picture is usually of a sample in a small, hard-to-reach location, fiber optic sensors may also be used in various other conditions (test tubes,

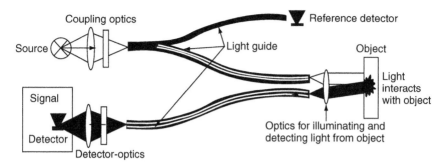

Figure 9.5 Configuration of a fiber optic sensor system.

microtube, tilted surface, moving mechanical part, etc.) [15]. Microbial biosensors may also incorporate transducers which convert biochemical signals into electrical or optical signals. Some sensors place immobilized micro-organisms close to a thermistor which measures the metabolic heat they produce; calorimetry may be applied to measure a large variety of substances based on thermal reactions such as this. Some sensors work in conjunction with microchips. For example, Oak Ridge National Laboratory has developed an infrared microspectrometer. It can be used for blood chemistry analysis, gasoline octane analysis, environmental monitoring, industrial process control, aircraft corrosion monitoring, and detection of chemical warfare agents. It uses a light source to excite chemicals in its environment, then channels the emitted light into an optical fiber for external analysis. The measured wavelengths are fed into a microchip, which identifies and determines the concentrations of chemicals in a sample. These miniaturized devices can be as accurate as standard laboratory procedure, with a large number of additional applications [17]. Finally, we should note that there are many other types of biosensors that use microchips, or illuminate using fluorescence, which do not require optical fibers.

References

[1] Hecht, J. (1999) City of Light: The story of Fiber Optics, Oxford University Press.
[2] Winawer, S.J., Fletcher, R.H., Miller, L., Godlee, F., Stolar, M.H., Mulrow, C.D., *et al.* (1997) "Colorectal Cancer Screening: Clinical Guidelines and Rationale". *Gastroenterology*, 112: 594–642 (Published erratum in Gastroenterology 1997;112:1422).
[3] Bongsoo, L. (2004) Fiberoptic Tutorial, Nanoptics Inc. May 6.

[4] Dennis C. Leiner (2002) Miniature Optics in the Hospital Operating Room, Lighthouse Imaging Corporation.

[5] Johnson, B.A. (1999) "Flexible Sigmoidoscopy: Screening for Colorectal Cancer", American Family Physician, January 15. American Academy of Family Physicians. p. 313.

[6] Reinhard, J. (2000) *Fundamentals of Fiber Optics (for Illumination)*, Volpi AG.

[7] Jerome D. Waye (2001) The Evolution of Gastrointestinal Endoscopy at The Mount Sinai Hospital, *The Mount Sinai Journal of Medicine*, 68 (2), March.

[8] Nathan, R., Annetine C. Gelijns, and Holly, D. (1995) *Sources of Medical Technology: Universities and Industry*, National Academy Press.

[9] Basil I. Hirschowitz (2000) "Endoscopy – 40 Years since Fiber Optics", *Digestive Surgery*, 17:115–117.

[10] Gregory, A., Michael, K., Harvey, W., Thomas Cox, J., Elliott, L., Jame McCaughan Jr., Brian, S., Raymond, L., *Overview of Clinical Laser Applications*, The Laser Training Institute, www.lasertraining.org/Clinical Laser Applications.PDF.

[11] For lasers in eye surgery, see http://www.lasersite.com/lasik/index.htm.

[12] For lasers in dental surgery, see http://www.dentalfind.com/laser_gum_surgery/.

[13] Website for Clark Equine Clinic, http://www.clarkequine.com/surgery.htm.

[14] Alison, P. (2004) "KTP laser shows high efficacy in men with retention", *Urology Times*, January 1.

[15] "Breakthrough Treatment for Enlarged Prostate", Surgical Center of Central Florida, website feature article, http://www.thesurgicalcenter.net/Feature.html.

[16] McLeish, T. (2000) "Devices Detect Salmonella, E. Coli, Other Bacteria", Newswise, June 26. University of Rhode Island, Kingston R. I.

[17] Bruce, J.K. (1996) "Biosensors and Other Medical and Environmental Probes", *ORNL Rev.*, 29 (3), Oak Ridge National Laboratory. http://www.ornl.gov/info/ornl-review/rev29_3/text/biosens.htm; also, see http://www.microfab.com/technology/biomedical/biomed.html.

Chapter 10 | Fiber Facts

This chapter, and the remaining chapters in this book, depart from the conventional approach used in the previous sections. As we have seen, there are a wide range of topics which are of interest to fiber optic practitioners, researchers, and design engineers. Frequently, it is convenient to have a list of "talking points" to include in presentations, reports, and classroom lectures; this section provides a sample list of "fiber facts" as a starting point. These facts should be taken in the context provided by the previous chapters. Details such as the number of phone calls which can be carried over a single fiber are likely to change over time, but can be easily modified based on updates to the references provided. The reader is encouraged to build their own list of facts based on their needs and applications.

10.1 Basic Information

- A single-mode fiber has a 8–10 micron diameter core. There are two standard multimode fibers: either a 50 micron diameter core or a 62.5 micron diameter core. There are other types of fiber available, though less commonly used for communication links; for example, a 100 micron core diameter can also be used for certain applications. The 50 and 62.5 micron core fibers both have a cladding of 125 microns diameter, and the 100 micron core fiber has a cladding of 140 microns. By comparison, a human hair is roughly 17–181 microns in diameter [1].

- In 2003, the total worldwide fiber market remained flat, at 55 million fiber km [2].

The equatorial circumference of the earth is 40,075 km (Wikipedia). This leads to a world fiber market that could be wrapped 1372 times around the equator of the earth. The average distance from the earth to the moon is 384,403 km (Wikipedia). This leads to a world fiber market that could travel 71.5 round trips to the moon.

- In 1998, one pair of fiber optic strands could transmit the equivalent of 129,000 simultaneous phone calls. In 2004, with transmission capabilities of 40 Gbit/s available over single-mode fiber, about 10,000,000 phone calls could be transmitted using one pair of fiber optic strands. Using dense wavelength division multiplexing, which allows 32–128 different wavelengths to be sent simultaneously, this can be increased to 320,000,000–1,280,000,000 phone calls.

The theoretical limit of the bandwidth of an optical fiber is 100 THz – this was reported within the past 10 years, prior to this time it was widely reported that the theoretical limit was about 25 THz. Since a phone call requires about 4 kHz, this leads to a theoretical upper limit of 2,500,000,000 phone calls on a single pair of fiber optic strands. This cannot be increased by multiplexing – it is a fundamental bandwidth limit of the fiber.

10.1.1 MOORE'S "LAW" AND OPTICAL DATA COMMUNICATION

The number of *transistors* per square inch on *integrated circuits* had doubled every year since the integrated circuit was invented [3].

The observation was made in 1965 by Gordon Moore, co-founder of *Intel* [4]. It is not a law of nature, but rather an empirical observation that has held up remarkably well since it was first suggested. For comparison, current IBM enterprise servers have about 50 million transistors on a 14.7 mm chip.

This has been one of the driving forces behind higher data rate input/output (I/O) channels for servers, based on fiber optic technology.

There has been a mismatch in the growth of processing power in computer central processing units (CPUs) and I/O bus speed. Between 1981 and 2001, CPU speed has increased from 5MHz (16 bits) to 1.2 GHz (64 bits), or 960 times. In the same time period, bus speed has increased from 8 MHz (8 bits) to 66 MHz (64 bits), or only 66 times [5].

Storage data density has been doubling annually since about 1997 (faster than Moore's Law). As a consequence, cost of storage has been steadily declining; we have gone from systems in the 1970s that were $60/MB of disk storage to current systems at 50 cents/MB of disk storage. Thus, fiber optic attachment to remote storage devices has also become a growth area.

10.1.2 INTERNET TRAFFIC

Internet traffic has been increasing at a rate between 70 and 150% (doubling) per year from 1996 to 2002. In 1996, Internet traffic was approximately 1500 TBytes/month. In 2002, it was between 80,000 and 140,000 TBytes/month.

There was a two-year period between 1995 and 1996 when the growth was much more rapid. The myth "Internet traffic doubles every ten weeks" came from that period. This statement lived longer than the data supporting it, because people began to confuse capacity with traffic by overestimating utilization. This made them believe there was a "insatiable demand for bandwidth," which was one factor that led to overbuilding of the fiber infrastructure and the telecom crash. Data through 2003 leads one to believe that the steady annual doubling of traffic is still in effect [6].

It has been proposed that there may be a Moore's Law for data traffic. However, since this type of law is not a law of nature, but instead an observation of behavior, it cannot necessarily be used predictively [7].

10.1.3 CAPACITY VS. COST

From 1960 to 1995 the cost performance of the leased lines required to make a cross country packet network were decreasing so slowly that they only halved every 79 months. DWDM in the mid-1990s resulted in a major decrease in the cost of long-haul communications estimated at a factor of two every 12 months [8].

A brief history of fiber optics in submarine networks [9]:

1857	Telegraph cable	Morse Code only, no voice traffic	Custom designed copper cables	Data rate N/A
1956	Transatlantic telephone cable	47 voice circuits	Coax cable and bidirectional regenerators	Data rate N/A
1980s		4,000 voice circuits	Improved cable and repeater technology	Data rate N/A
1988	TAT-8	35,000 voice circuits	Hybrid optical system	280 Mbits/sec
1995–1996	TAT-12/13		Fully optical system with optical amplifiers and no high speed electronics in submerged repeaters	5 Gbits/sec
2002	i2iCN		Eight fiber pairs with DWDM and wideband erbium doped fiber amplifiers	8.4 Tbits/sec

10.2   Disaster Recovery Facts

Fiber optic networks played an important role in disaster recovery follow-ing the September 11, 2001 terrorist attacks in New York City. To better appreciate the role played by optical networks, consider the following facts [10].

Here is how (Bill) Moore (telecommunications manager at the Museum of Modern Art in midtown Manhattan, and an officer with the Commu-nications Managers Association) described the communications infras-tructure in lower Manhattan: "(It) is a maze of tunnels. Most of the (fiber-optic) cables run below ground from building to building in those tunnels. That area down there is where Manhattan narrows to a point, and everything converges down there. As the World Trade Center buildings collapsed, they collapsed down into those tunnels. And as they start doing the excavation down there, there are likely to be some tractor cuts, too."

Wall Street's main interconnection point is the heavily damaged Verizon West Street facility, which supported 200,000 voice lines and 3 million private lines prior to the September 11 attacks.

Verizon had no easy answers to questions about why there was not more alternate routing. Despite using highly reliable technology such as SONET, the reality is that even SONET's dual-ring architecture cannot help when both the fiber rings and the equipment attached to them are ruined.

Within a few days after the disaster (Verizon) had rerouted 24 OC-48 circuits, which is the equivalent to more than 200 million T-1 lines, according to (Ivan) Seidenberg (co-CEO and president of Verizon).

AT&T lost landline connectivity to its wireless cell sites but reported that its local switching gear located several stories down in the basement of the World Trade Center remained unscathed. Still, AT&T says it rerouted calls. Qwest Communications and WorldCom say their networks were not affected by the attacks.

Disaster-recovery vendor SunGuard Recovery Services disclosed that 15 customers have been affected so severely as to move into SunGuard sites, where they are provided with full telecommunications and computer equipment to duplicate their office needs. Another 70 customers alerted SunGuard that they may need to use its services. Separately, disaster-recovery service provider Comdisco says 32 customers declared disaster emergencies.

SoundView (Technologies Group, an online investment banking firm) has dedicated, private voice lines—called ring-down circuits—at all the major brokerage firms in New York, and most of those lines were down. "With the ring-down circuits, you just pick up a line, and you are talking to the broker," explains Bart Anthony Lavore, manager of corporate telecommunications for SoundView. "These are voice lines that can be copper or (optical) T-1 circuits that go straight to the brokerage house" [11].

Disaster-recovery specialist Comdisco Inc. says 48 New York businesses moved into its backup facilities in Queens, NY, and in two New Jersey locations in the hours after the terrorist attacks.

EMC had 76 customers in and around the World Trade Center, and the largest and most technically sophisticated were protected by investments they had made in real-time data backup at secondary data centers.

The telecommunications infrastructure in New York was severely strained. Verizon's switching center at 140 West St., which served as a major wiring hub for the World Trade Center, was heavily damaged, and the phone company's Broad Street office, which provides 80% of the data circuits serving the New York Stock Exchange, lost electricity.

Verizon's West Street central office next to the World Trade Center supported 200,000 voice lines and 3 million private lines before September 11, and its Broad Street office handled 80% of the New York Stock Exchange and Nasdaq telecom needs.

For Verizon, it took 4 million rebuilt or rerouted voice and data circuits, 3,000 technicians and managers, and 18 new SONET rings to get its network in lower Manhattan back in order. All was done in two months.

Verizon has installed better quality inter-central-office fiber to replace the links destroyed at its damaged site at 140 West St., which had a hole punctured in its wall when the World Trade Center building collapsed on the central office. And in buildings where Verizon engineers determine there will be enough traffic demand, the carrier is replacing copper connections with fiber.

"If customers in those buildings wanted to upgrade from multiple T-1s to a T3, we could accommodate that now," according to Joe DeMauro, Verizon's regional president for cabling in the affected area.

On September 11, the hospital (NYU Downtown Hospital), which had redundant connections, lost its data and voice service when the two central offices to which it was linked—one in the World Trade Center and the other at 140 West St.—were knocked out. Within four days, Verizon and AT&T had laid cable in the street to reconnect the hospital (despite the two companies usually being competitors). Several circuits from MCI, which were supposed to provide some redundancy, all ran into a central office at the World Trade Center. When that central office went down, the hospitals lost all their MCI circuits, which are now being replaced with lines from Verizon and AT&T that run into multiple central offices.

Since September 11, Verizon has restored 100,000 business voice lines, 200,000 residential voice lines, 3.6 million data circuits, and 10 cellular towers [12].

10.3   Fiber for Lighting and Illumination

As early as the 1800s, it was known that water jets and bent glass rods could guide light. By the 1880s, water fountains all over America and Europe were being illuminated.

Fibers used for illumination need to carry visible wavelengths (between about 400 nm in the blue and around 750 nm in the red), not infrared light as in communication networks. Thus, incandescent lamps such as halogen or metal halide are often used. Fibers used in lighting systems have much larger diameters (a fraction of a mm to several mm). Unlike communication systems, lighting systems may deliberately cause light to leak out the sides of a fiber as an alternative to decorative neon tubes. Fiber lighting systems offer several advantages; the light source can be placed in a convenient location for access and maintenance, and one source can be used to light several areas. With the correct filters, lighting fibers will not transmit infrared or ultraviolet light, making them well suited for museums and showcase lighting since they will not damage paintings, textiles, or other sensitive materials. Glass or plastic fibers are also nonconductive, making them good choices for wet environments such as gardens or swimming pools. Care must be taken with plastic fibers, which may tend to turn yellow in color over time [13].

Fiber optic lighting is used for various decorative effects, including lighting walkways, landscaping, countertops, and stairways. Starlight patterns on ceilings can be produced using one or two light sources to light several dozen fiber tips. Decorative tree lighting is another example of optical fiber illumination systems.

In "hybrid" lighting systems for buildings, sunlight is gathered on the roof and piped through optical fibers, so rooms have a combination of natural and artificial light. Since the light from the fiber optic system does not emit heat, air conditioning costs are reduced. Significant savings in electricity are also possible; for example, in the U.S. about 25% of the electricity generated is used for lighting, at a cost of over $100 million per day; hybrid lighting could reduce this by up to one-third [14].

Fibers can also be used for backlighting liquid crystal displays (LCDs). For example, individual fibers can be woven like threads into a single layer; the bends created by the weaving are controlled to slightly exceed the bend radius of the fibers, allowing some light to escape. In another approach, the fiber core is abraded or scratched in a controlled way to allow light to escape from the core. The fibers can then be bundled behind an LCD panel and routed to a remote light source.

Fibers can also be used for automotive lighting. In one approach, a lamp system is installed in the car's trunk and fibers are used to carry light to the front of the car; this is the basis for the adaptive front lighting system (AFS), which can turn the headlights to increase lighting around curves [15].

Plastic optical fiber networks are being installed in luxury cars, such as the Renault Espace; this channels DVD movies, digital TV, video games, or GPS navigation to individual LCD screens mounted throughout the car, as well as a parking camera which produces a view from behind the car on a dashboard screen. The system is based on plastic fiber from Yazaki Optical Cable Co. and 250 Mbit/s red resonant-cavity LED transceivers [16].

The following are some websites with examples of fiber optic lighting.

- Fiber with side emissions—http://www.fiberopticproducts.com/Novabright.htm

- LCD backlighting—http://www.lumitex.com/index.html

- Automobile lighting
  - http://www.us.schott.com/fiberoptics/english/products/automotive/lighting/index.html
  - http://www.lrc.rpi.edu/programs/Futures/LF-Auto/interior.asp

- Landscaping and architectural lightning—http://www.digilin.com.au/applications.htm

- Ambient and task lighting—http://www.us.schott.com/fiberoptics/english/products/lighting/applications/home.html

- Woven optical fibers—http://www.lumitex.com/pdfs/Brochure.pdf

- Abraded optical fibers—http://www.lumitex.com/pdfs/Brochure.pdf

- Decorative lighting
  - http://www.glassblocklighting.com/
  - http://www.digilin.com.au/applications.htm
  - http://www.ez-tree.com/scripts/prodView.asp?idproduct=617
  - http://www.us.schott.com/fiberoptics/english/products/lighting/applications/home.html
  - http://www.fiberopticproducts.com/Ceiling.htm
  - http://www.digilin.com.au/Residential%20Domestic.htm
  - http://www.fiberopticproducts.com/Novabright.htm

- Lighting a building
  - http://www.ornl.gov/info/ornlreview/rev29_3/text/hybrid.htm

- Lighting swimming pools—http://www.digilin.com.au/Residential%20Domestic.htm

10.4   Unusual Applications and Future Developments

- Fiber optic medical fact—Prostate enlargement is a condition that causes urinary problems in nearly half of all men over the age of 50, and that percentage increases with age. By the age of 80, nearly 80% of all men suffer from BPH symptoms. The GreenLight PVP Procedure uses a small flexible fiber optic into the urethra, and the fiber vaporizes the enlarged prostate tissue. Typically, this is performed as an outpatient procedure. TURP, the traditional surgical alternative, requires long hospital stays, increased pain, and risk of complication and death in elderly patients.

- Supercomputers have begun to use parallel optical fibers to cluster more processors and memory together at higher bandwidths. This could significantly increase the power of next generation computing systems. While it is difficult to compare the computing power of a machine to the human brain, various estimates (such as the density of retinal cells extrapolated up to the volume of the brain) suggest that the compute power of a human brain will be available to clustered supercomputers as early as the year 2015 [17].

- An anti-tank missile uses fiber optic cable for flight control. Signals on fiber optic cables cannot be jammed. These are called tow missiles, because they have a tow rope.

- Fiber optic sensors have a variety of non-communications applications. For example, an intrusion alarm system uses fiber optic cable as a sensing element.

- Non-traditional fiber growth techniques—Deep-sea sponges grow fibers for anchorage and structural support that are both stronger than commercial fibers and exhibit higher light transmission [18].

- Nanofibers (silica waveguides with diameters smaller that the transmitted light) have been demonstrated in the laboratory. Light can be launched by optical evanescent coupling into fibers that can be fabricated as small as 50 nm (0.05 microns). The critical diameter for single mode propagation is 450 nm (0.45 microns) at 633 nm wavelengths, and 1100 nm (1.1 micron) at 1550 nm wavelengths. About 20% of the light propagates outside as an evanescent wave for the core at these sizes, and for smaller sizes the percentage goes up. These devices may be used for future microphotonic devices [19].

- Solitons, which are often mentioned in science fiction, are a technique of creating ultra-short pulses that propagate through the fiber without changing. Optical solitons are a non-linear solution to the wave equation that balances dispersion and non-linearity, and thus they do not degrade the signal quality with the propagation distance. Soliton pulses propagate through the fiber without changing their spectral and temporal shapes.

Solitons are one of the few nonlinear optical effects that solves a practical problem. Most nonlinear optical effects, such as Raman scattering, are sources of noise. The key factors that have limited the practical application of solitons to fiber optics have been that they are difficult to establish and maintain, and it is now possible for a conventional fiber optic signal to reach halfway around the world without requiring this technology.

- Faster than light (FTL)—Highly controversial FTL signal transmission experiments using stimulated emission of negative index materials could open up link distances and transmission speeds always thought to be prohibited by special relativity. A research group in Princeton appears to have sent narrow optical pulses through excited atomic cesium vapor at superluminal speeds, and various other experiments suggest that FTL propagation of optical and microwave signals is possible.

Further Reading

DeCusatis, C. (2001) *Analog Magazine*, CXXI (9) pp. 44–53, September.

Margos, J. (2000) *Nature*, 406, pp. 243–244, July 20.

Mugnai, D., Ranfagni, A., and Ruggeri, R. (2000) Physical Review Letters, 84(21), pp. 4830–4833, May 22.

Wang, L.J., Kuzmich, A., and Dogariu, A. (2000) Nature, 406, pp. 277–279, July 20.

References

[1] The Physics Factbook website, http://hypertextbook.com/facts/, entry by Brian Ley, 1999.

[2] Corning OFC speech, 2004.

[3] http://www.webopedia.com/TERM/M/transistor.html,   http://www.webopedia.com/TERM/M/integrated_circuit_IC.html.

[4] http://www.webopedia.com/TERM/M/Intel.html.

[5] William T. Futral, InfiniBands Architecture Development and Deployment, p. 9.

[6] Internet traffic growth: Sources and Implications by A.M. Odlyzko.

[7] Internet growth: Is there a Moore's Law for data traffic? K.G. Coffman and A.M. Odlyzko, July 11, 2000.

[8] Paraphrased from Internet Growth Trends, Dr. Lawrence G. Roberts, Chairman, Packetcom, http://www.ziplink.net/~lroberts/IEEEGrowthTrends/IEEEComputer12-99.htm.

[9] Vincent, L. (2004) "Submarine systems: from laboratory to seabed", *Optics & Photonics News*, February, 15(2).

[10] "Internet, telecom networks put to test in wake of terrorist strikes on U.S.", Network World, September 17, 2001.

[11] John, F. (2001) "Ready for Anything", *Information Week*, September 24.

[12] Michael, M. and Denise, P. (2002) "Carriers Stay Course with NYC Networks", *Network World*, March 11.

[13] Knisley, J.R. (2005) Fiber Optic Lighting Fundamentals, http://ewweb.com/mag/electric_fiberoptic_lighting_fundamentals/, April 22.

[14] Rist, C. (2005) "An able cable", This Old House, http://www.thisoldhouse.com/toh/print/0,17071,198683,00.html, April; also see Cates, M.R. (2005) Hybrid Lighting: Illuminating Our Future, http://www.ornl.gov/info/ornlreview/rev29_3/text/hybrid.htm.

[15] "The long and lighted road", RPI Lighting Research Center, http://www.lrc.rpi.edu.

[16] Graydon, O. (2005) POF Network Offers In-car Entertainment, http://www.fibers.org, 26 April.

[17] Advanced Strategic Computing Initiative (ASCI) roadmaps, http://www.llnl.gov/asci; Kurzwell, D. (1999) The Age of Spiritual Machines.

[18] Lucent, optics.org. For more information on Lucent, visit http://www.lucent.com. For more information on their Deep Sea Sponge Research, visit http://www.lucent.com/press/0803/030821.bla.html.113.

[19] Tong, L., Gattas, R.R., Jonathan B. Ashcom, He, S., Lou, J., Shen, M., Maxwell, I., and Mazur, E. (2003) "Subwavelength-diameter silica wires for low loss optical waveguiding", *Nature*, 426:816.

Chapter 11 | Fiber Optic Timeline

Fiber optics is often regarded as a recent development, perhaps because of its role in the backbone of the Internet. Actually, the historical roots of this technology stretch back to the Victorian era; it has only been within the past few decades that diverse technologies such as high speed electronics, semiconductor solid state lasers, and low loss glass optical fibers have come together to form the fundamentals of modern optical communication networks. There are several excellent references available which describe the early development of this field in detail; for our purposes, we will only note a few significant milestones along the way. The concept of a global communications network, with all the associated concerns about privacy, security, taxation, and electronic marketing scams, dates back to the development of telegraph networks in the 1800s [1]. The notion of using light as a communication medium may have been adapted from even earlier signaling methods using bonfires, smoke signals, mirrors, or telegraphs with movable arms; indeed, light guiding through water was a well established phenomena at this time [2], although it was considered little more than a parlor trick with no practical applications. Following the development of the telephone, in 1880 Alexander Graham Bell proposed a wireless communications technology using modulated sunlight which he dubbed the photophone [3].

Despite the lack of commercial success for this product, the concept of optical communication and the tools to make it a reality continued to be pursued by many early pioneers in the field. All of the basic concepts for optical fiber communication were outlines by C.W. Hansell as early as 1927, in a patent designed to transmit images over bundles of glass fiber from remote locations [4]. The concept of transmitting multiple wavelengths is also nothing new, having its origins as early as the 1930s in a series of experiments by a French scientist dealing with propagation of light through mirrored tubes [5]. The basic idea of transmitting multiple colors of light appears obvious to anyone observing rainbows; modern wavelength multiplexing has an analogy in the form of rainbows of infrared light, which not only exist in nature but can be made visible and photographed using various techniques [6]. With the development of low loss, single-mode optical fibers in the late 1970s, the potential for applying this scheme to practical communication systems began to emerge. It was around this time that Will Hicks put together some of the earliest concepts behind wavelength multiplexing; at the time, he was criticized for his vision that a single optical fiber might carry hundreds or more different wavelengths of light, and that one optical amplifier might be able to simultaneously boost the strength of all these signals [7]. Today, of course, this technology has become a reality which is used in every major telecommunication and data communication network worldwide.

There are many ways to measure the growth of optical networks. A study by the National Science Foundation, for example, has suggested that the bit rate—distance product for WDM systems—has grown steadily over the past decade or so, approaching 10e7 Gbit/s-km in the year 2000. Similarly, this same study concluded that the total capacity of optical networks using TDM has increased from roughly 1 Gbit/s to a bit over 10 Gbit/s in the past 10 years, whereas the capacity of DWDM has gone up two orders of magnitude (1 Gbit/s to over 100 Gbit/s) in half the time (a five-year period from 1995 to 2000).

We refer the interested reader to an excellent history of optics, including a list of the major people involved in its development and a chronology until 1996, provided in reference [7]. In this chapter, we will supplement this chronology through the following timelines:

1. Basic knowledge timeline—A short general timeline.
2. Timeline for selected optical data communications milestones.
3. Timeline for fiber optic milestones in medical applications.

11.1   Basic Knowledge Timeline

2500 BC–1700s	Glass manufacture is developed, including drawing glass into fibers.
1837	Introduction of Morse Code and the telegraph.

1840– 1889	Light guiding is proposed, and used in water jets as a decoration at several public displays (Paris opera—1853, International Health Exhibition, London—1884, Universal Exhibition in Paris—1889).
1880	Alexander Graham Bell invents the photophone, a means of communicating voice signals using modulated light.
1888	First medical applications of Fiber Optics—bent glass rods to illuminate body cavities for dentistry and surgery.
1900	Early theoretical papers concerning waveguides for light begin to emerge.
1926	Birth of Fiber Optic Imaging—in that one year, several different television scheme patents were filed, including one that proposes a fiber optic imaging bundle (Clarence Hansell).
1942	Invention of the German "Lichtsprecher," a simple optical communication device.
1951	Cladding is proposed independently by Holger Moller Hansen in Denmark and Brian O'Brien at the University of Rochester (O'Brien suggests the idea to Abraham van Heel, who makes the first clad fiber bundles).
1956	Curtiss makes first glass-clad fibers by rod in tube method.
1960	Theodore Maiman demonstrates the first laser.
1961	Eli Snitzer of American Optical publishes theoretical descriptions of single-mode fiber.
1962	Four groups develop first semiconductor diode lasers which emit pulses at liquid nitrogen temperatures.
1963	Heterostructures proposed for semiconductor lasers.
1964	Charles Koester and Eli Snitzer describe first optical amplifier, using neodymium doped glass.
1967	Shojiro Kawakami of Tohoku University in Japan proposes graded-index optical fibers.
1969	Martine Chown of Standard Telecommunication Labs demonstrates fiber optic repeater at Physical Society exhibition.
1970	First continuous wave room temperature semiconductor lasers made in early May by Zhores Alferov's group at the Ioffe Physical Institute in Lenningrad (now St. Petersburg) and on June 1 by Mort Panish and Izuo Hayashi at Bell Labs.

1970   Maurer, Donald Keck and Peter Schultz at Corning make a single-mode fiber with loss of 16 dB/km at 633 nm by doping titanium into fiber core.

1973   John MacChesney develops modified chemical vapor deposition process for making fiber at Bell Labs.

1974   Diode laser lifetime reaches 1000 hours at Bell Labs.

1976   Masahara Horiguchi (NTT Ibaraki Lab) and Hiroshi Osanai (Fugikura Cable) make first fibers with low loss—0.47 dB/km at long wavelengths.

1976   Lifetime of best laboratory lasers at Bell Labs reaches 100,000 hours (10 years) at room temperature.

1976   J. Jim Hsieh makes indium gallium arsenide phosphide (InGaAsP) lasers emitting continuously at 1.25 micrometers.

1977   Several telephone companies start using fiber optics for normal traffic. (AT&T—Chicago's Loop district, General Telephone and Electronics, Long Beach California, Bell System, downtown Chicago and British Post Office, Martlesham Heath).

1977   Bell Labs announces one million hour (100 year) extrapolated lifetime for diode lasers.

1977   AT&T and other telephone companies settle on 850 nm GaAs light sources and graded-index fibers for commercial systems operating at 45 million bits per second.

1980   AT&T asks Federal Communications Commission to approve Northeast Corridor system from Boston to Washington through graded-index fiber.

1982   MCI purchases right of way to install single-mode fiber from New York to Washington.

1988   TAT-8, first transatlantic fiber optic cable, begins service using 1.3 micrometer laser and single-mode fiber.

1991–   Masataka Nakazawa of NTT reports sending soliton signals
1993   through a million kilometers fiber in 191, and through 180 million kilometers in 1993.

1995   Invention of first long wavelength (1.55 micrometer) vertical cavity surface emitting laser (VCSEL) by a team led by Dr. Dubravko Babic at Hewlett Packard.

1995–   TAT-12/13 first fully optical submarine system laid across the
1996   North Atlantic Ocean. Each fiber carried 5 Gbits/s.

2000   Invention of 1.3 micrometer vertical cavity emitting laser
         (VCSEL) by Sandia National Laboratory, working with Cielo
         Communications Inc.

2002   i2iCN submarine cable, which links Singapore to India,
         designed to carry up to 8.4 Tbits/s by the use of 10 Gbits/s
         DWDM technology.

11.2 Timeline for Selected Optical Data Communications Milestones

1982   ANSI X3T9.5 committee proposal for a high speed token pass-
         ing ring as a back-end interface for storage devices (would later
         become FDDI).

1985   Discovery of erbium-doped fiber amplification by S.B. Poole,
         University of Southhampton, England (from a concept proposed
         as early as 1961).

1988   Development of Fibre Channel Standard specifications begins
         for 1, 2, 4, and 10 Gbit/s data rates.

1988   Release of SONET standards for optical links that will come
         into widespread use across the datacom and telecom industry.

1990   Introduction of multimode ESCON channels on IBM enter-
         prise servers, including the definition of the retractable shroud
         ESCON duplex connector.

1991   Bell Labs demonstrates the first erbium doped fiber amplifier
         (EDFA).

1992   Introduction of single-mode ESCON channels. IBM releases
         Parallel Sysplex architecture, including fiber optic links to dis-
         tribute time of day clock signals and for clustering of enterprise
         servers. FDDI standard approved.

1993   Formation of the modern ITU from the former CCITT standards
         body.

1994   ANSI Fibre Channel Standard released for 1, 2, 4, and 10 Gbit/s data rates. Early work on multi-fiber MPO-type optical connectors.

1996   Release of industry standardized ESCON links, known as SBCON. Commercial release of the first ten-channel dense optical wavelength division multiplexing for data communications (IBM 9729 optical wavelength division multiplexer).

1997   Release of updated Ethernet specifications, IEEE 802.3 family of standards, including optical links near 1 and 10 Gbit/s data rates. Support for Geographically Dispersed Parallel Sysplex on the IBM 9729 WDM. Development of the IBM FICON interface specifications.

1998   Formation of small form factor (SFF) optical interface groups and multi-source agreements for optical transceivers and cables. Drafting of serial HIPPI standard serial fiber optic interface. Drafting of initial gigabit Ethernet specification.

1999   Commercial release of the IBM 2029 Fiber Saver, the first 32 wavelength multiplexing system for data communications.

2000   Release of initial InfiniBand specification, including parallel optical links at 2.5 Gbit/s/line for 1X, 4X, and 12X widths. Formation of pluggable small form factor (SFP) optical transceiver multi-source agreements. Global consumption of fiber optic cable passes $15.8 billion.

2001   Large scale demonstration of disaster recovery using optical networks during terrorist attacks in New York City and Washington, D.C. Initial release of optical transmitter/receiver arrays with greater than 12 fibers per interface. Initial development of 10 Gbit/s Ethernet standards.

2002   Advanced Supercomputing Initiative (ASCI) program delivers first working samples of parallel optical links in technical supercomputers. Formation of the XFP multi-source agreement for 10 Gbit optical transceiver packaging. Development work on the Generic Frame Procedure standard for protocol encapsulation in SONET.

2003   Commercial use of parallel optical links for high-end technical computer clustering (12X wide, 2 Gbit/s/line). EIA/TIA standards proposal to increase fiber count in multi-fiber connectors up to 96 fibers per MPO.

2004	Release of modified Fibre Channel specs for 8 Gbit/s data rate. Release of updated InfiniBand specification, including active optical cable assemblies, 8X line widths, 5 Gbit/s data rates for serial and parallel optical links, 10 Gbit/s data rate for serial optical links, and pluggable parallel optical transceivers. First nanoscale optical fiber waveguides capable of mass production, Prof. Metzer, Harvard. Multi-source agreement for pluggable small form factor transceivers in WDM applications announced.
2004–2005	Practical long wavelength (1300 nm) VCSEL sources begin to emerge. 4 Gbit/s Fibre Channel links begin to emerge as common storage area networking interconnects.
	Discussions under way for 16 Gbit/s Fibre Channel, as well as 40 Gbit/s and 100 Gbit/s Ethernet interfaces. Ongoing research proposals for massively parallel optical interconnects in clustered computing environments.

11.3 Timeline for Fiber Optic Milestones in Medical Applications

1879	Nitze built a 7 mm cystoscope with burning platinum wire illumination [4].
1888	Dr. Roth and Prof. Reuss of Vienna use bent glass rods to illuminate body cavities for dentistry and surgery [2].
1898	David D. Smith of Indianapolis applies for patent on bent glass rod as a surgical lamp [2].
1901	Rigid endoscope (not using fiber optics) used on dog [5].
1920s	Bent glass rods common for microscope illuminations [2].
December 30, 1926	Clarence W. Hansell proposes a fiber optic imaging bundle in his notebook at the RCA Rocky Point Laboratory on Long Island. He later receives American and British patents [2].
1930	Heinrich Lamm, a medical student, assembles first bundle of transparent fibers to carry an image (of an

	electric lamp filament) in Munich. He does not get the patent because of Hansell's prior work [2].
1936	Schindler built a semi-flexible gastroscope—48 lenses in spiral spring [4].
1939	Curvlite Sales offers illuminated tongue depressor and dental illuminators made of Lucite, a transparent plastic invented by DuPont [2].
1952	Harold Horace Hopkins applies for a grant from the Royal society to develop bundles of glass fibers for use as an endoscope at Imperial College of Science and Technology. Hires Narinder S. Kapany as an assistant after her receives the grant [2].
1952	Fourestier, Gladu, and Vulmiere provided the first description of gynecological fiberoptic endoscope with light transmitted by optical fiber. Device was developed by French gynecological surgeon Palmer in conjunction with Richard Wolfe optics in Germany. Karl Storz introduced its "cold-light" endoscope a few years later in 1960. Prior to this, attempts at surgical contraception using rigid endoscopes involved incision, and possible burns by the lamp at the tip [5].
February 18, 1957	Hischowitz tests first fiber optic endoscope in a patient [2, 4, 5].
1960	Hopkins files Rod lens patent [4].
1961	First commercially available fiber optic gastroscope, American Cystoscope Makers Inc (ACMI) 4990 Hirschowitz fiberscope [5].
1965	Fiber optic improvement to pre-existing gastrocamera (1955). The gastrocamera was a miniature camera attached to the tip of an endoscope. But as fiberscopes improved, "it soon thereafter became much easier for most endoscopists to simply attach an external 35 mm camera to the eyepiece of the gastroscope and photograph the image conveyed by the fiber bundle" [5].
Late 1960s	Fiber optic endoscope refinements—better optical clarity, easier to manipulate distal tip, adding channels for biopsy, and therapeutic maneuvers [5].

1966   Hopkins endoscope introduced by Storz. Improved rigid endoscope—major increase in light transmission and wider viewing angle [5].

At this point, gastroenterology became firmly committed to flexible fiber optic technology, but gynecological endoscopy was still improving and miniaturizing rigid endoscopy. These rigid endoscopes used fiber optic illuminators. The advantage of rigid endoscope technology was increased image clarity, because they were not using a fiber bundle [5].

1967   Bergein Overholt, from the academic medical center of the University of Michigan, reports to the meeting of the American Society for Gastrointestinal Endoscopy, the first examinations using fiber optic colonoscopy. Before fiber optics, rigid, lens-and-prism endoscopes could not reach through the sharp curvature of the sigmoid colon. Proctosigmoidoscopes could inspect only 25–30 cm of the proctosigmoid. With fiber optic colonoscopes, polypectomy, the removal of colon polyps, soon followed [5].

1967–   "The next decade saw the establishment of diagnostic
1977   endoscopy of both upper and lower GI tracts, and with the ability to biopsy then solidly supporting and refining visual diagnosis, diseases of the esophagus, stomach, duodenum, and colon began to be better diagnosed and their natural history understood" [7].

Expansion to a working length of 75 cm; the esophagoscope—an end viewing modification of the Hirschowitz fiberscope—became the first fiber optic endoscope to achieve widespread use [5].

Expansion to a working length of 110 cm tool now able to study esophagus, stomach, and duodenum in same procedure [5].

1968   Nippon Sheet Glass developed Selfoc (type of plastic lens used for endoscopes) [4].

1970   Endoscopic Retrograde cholangiopancreatography (ERCP) by Olympus and Machida in Japan. This technology is used for retained bile duct stones. In 1974, gastroenterologists from clinics in Germany and Japan extended the use of ERCP from diagnosis to therapy [5].

1970   Watanabe developed 1.7 mm gradient index endoscope [4].

Late   Video guided endoscopy [5].
1970s

1972   Gynecologist Wheeless designed and perfected the operating laparoscope, which was a rigid endoscope which allowed a physician to view as well as obstruct the fallopian tubes with one instrument, prevented the need for a second puncture [5].

In the 1980s ultrathin optical fibers were produced and as a result, the diameter of an endoscope could be reduced to less than 1 mm [6].

1983   Welch-Allyn replaces fiberscope with CCD at distal tip of endo-scope, creating an integrated video-endoscopy system. (The CCD was invented in 1969, and was not patented, making it available to all.) [5]

1987   Philippe Mouret in Lons France performed first human laparo-scopic cholecystectomy (gallbladder removal) using a gyneco-logical instrument [5].

1989   Laser fibers replaced electrocautery devices in laparoscopic cholecystectomy. Laser companies, such as Coherent, Trime-dyne, and Lasersope developed argon, dual wavelength KTP, and contact YAG lasers for use in laparoscopic cholecystec-tomy. FDA approval is sought [5].

1991   More than 50%, 32,750, of the United States practicing gen-eral surgeons learned laparoscopic cholecystectomy during the 18 months after the procedure was introduced. Gallblad-der removal is expanded from sicker to mildly symptomatic patients, as well as to higher risk patients once considered ineligible for the procedure. In Europe, the adoption of the procedure is roughly half that of the United States, partially because European manufacturers focus of the U.S. market, and partially on different payment mechanisms [5].

1998   The first laser is approved for LASIK eye surgery [8].

2001   GreenLight PVP Laser Procedure treatment of enlarged prostate receives FDA clearance in May 2001 [9, 10].

References

[1] Standage, T. (1998) **The Victorian Internet**, Berkley Books.
[2] Hecht, J. (1999) City of Light, Oxford University Press.
[3] Vincent, L. (2004) "Submarine Systems: From Laboratory to Seabed", Optics & Photonics News, February, 15 (2): 30–35.

[4] Dennis C. Leiner (2002) Miniature Optics in the Hospital Operating Room, Lighthouse Imaging Corporation.

[5] Sources of Medical Technology: Universities and Industry, Committee on Technological Innovation in Medicine, Institute of Medicine, 1995.

[6] Bongsoo, L. (2004) Fiberoptic Tutorial, Nanoptics Inc. May 6.

[7] Basil I. Hirschowitz (2000) "Endoscopy – 40 years since Fiber Optics: Any Light at the End of the Tunnel?", *Digestive Surgery*:115–117.

[8] "50 Years of SPIE".

[9] Alison, P. (2004) "KTP laser shows high efficacy in men with retention", *Urology Times*, January 1.

[10] *Breakthrough Treatment for Enlarged Prostate*, Surgical Center of Central Florida, website feature article, http://www.thesurgicalcenter.net/Feature.html.

Appendix A | Glossary

absorption   the loss of light when passing through every material, due to conversion to other energy forms, such as heat.

acousto-optic tunable filter   a light filter tuned by acoustic (sound) waves. This is accomplished using polarizers and an acoustic diffraction grating which creates a resonant structure that rotates the polarization.

active region   the area in a semiconductor which either absorbs or emits radiation.

aggregate capacity   a measure of the total information handling capability of a smart pixel array. It combines the individual channel data rate with the total number of channels in the array to produce an aggregate information carrying capacity.

alignment   the connection of optical components to maximize signal transmitted.

alignment sleeve   part of an ESCON connector into which the ferrules are inserted, assuring accurate alignment.

annealing   the process of heating and slowly cooling a material. This makes glass and metal stabilize its optical, thermal, and electric properties, and can reverse lattice damage from doping to semiconductors.

architecture   overall structure of a computer system, including the relationship between internal and external components.

asynchronous a form of data transmission where the time that each character, or block of characters, starts is arbitrary. Asynchronous data has a start bit and a stop bit, since there is no regular time interval between transmissions, and no common clock reference across the system.

attenuation the decrease in signal strength caused by absorption and scattering. The power or amplitude loss is often measured in dB.

Auger nonradiative recombination when an electron and hole recombine, and then pass excess energy and momentum into another electron or hole. This process does not generate any additional radiation.

backbone network a primary conduit for traffic that is often both coming from, and going to, other networks.

backplane circuit board with sockets to connect other cards, especially communication channels.

band gap the energy difference between the top of the valence band and the bottom of the conduction band of a solid. The size of the band gap will determine whether a photon will eject a valence electron from a semiconductor.

bandwidth the range of frequencies over which a fiber optic medium or device can transmit data. This range, expressed in hertz, is the difference between the highest and the lowest frequencies for optical filter elements. For MM fibers, the range is expressed as a product of the bandwidth and distance, in MHz-km.

baud the signaling speed, as measured by the maximum number of times per second that the state of the signal can change. Often a "signal event" is simply the transmission of a bit. Baud rate is measured in sec^{-1}.

beacon process a token ring process which signals all remaining stations that a significant problem has occurred, and provides restorative support. Like a string of signal fires, the beacon signal is passed from neighbor to neighbor on the ring.

beating when superimposing waves of different frequencies, beating occurs when the maximum amplitude of these waves match up. Coherent beating occurs when the data signal beats with itself out of phase, and can cause cross-talk.

biasing applying a voltage or a current across a junction detector. This changes the mode of the detector, which can effect properties such as the noise and speed. See photoconductive and photovoltaic mode.

bit   either 0 or 1, the smallest unit of digital communications.

bit error rate   the probability of a transmitted bit error. It is calculated by the ratio of incorrectly transmitted bits to total transmitted bits.

bit synchronization   when a receiver is delivering retimed serial data at the required BER.

Bragg reflector   see distributed Bragg reflector.

Brillouin scattering   the scattering of photons (light) by acoustic phonons (sound waves). A special case of Raman scattering.

broadcast-and-select network   a network that has several nodes connected in a star topology (see topology).

bulkhead splice, splice bushing   a unit that allows two cables with unlike connectors to mate.

burn-in   the powering of a product before field operation, to test it and stabilize its characteristics.

bus   see optical bus.

bus topology   see topology.

butt coupling   a mechanical splice between a fiber and a device. This technique is done with the signal on, and the fiber secured into position when measured signal is maximized (see splice).

butterfly package   see dual-in-line pin package.

byte   eight bits, numbered 0 to 7 (see octet).

cable jacket   the outer material that surrounds and protects the optical fibers in a cable.

cable plant   passive communications elements, such as fiber, connectors and splices, located between the transmitter and receiver.

calibration   the comparison to a standard or a specification. Also to set a device to match a specification.

channel   a single communications path or the signal sent over that path.

"chicken and egg" syndrome   when it is hard to find someone to take the first risk on a new technology, which delays its implementation. For example, when customers do not want to commit to using an emerging technology without some assurance that it will be fully supported by manufacturing, and suppliers do not want to commit to putting emerging technology into production without having a customer who is committed to using it.

chirp   a pulsed signal whose frequency lowers during the pulse.

cholesteric liquid crystals   see liquid crystals.

chromatic dispersion   see dispersion.

cladding   the material of low refractive index used to cover an optical fiber which reflects escaping light back into the core, as well as strengthens the fiber.

cleaving   to split with a sharp instrument along the natural division in the crystal lattice.

clock recovery   reconstructing the order of operations after serial/parallel conversion.

client   a computer or program that can download data or request a service over the network from the server (see server).

coherent beating   see beating.

correlators   a device that detects signal from noise by computing correlation functions, which is similar to transforms.

cost of goods sold   the sum manufacturing cost, serviceability and warranty cost, obsolescence and scrap and a portion of the supply chain cost (see parts cost, manufacturing cost, supply chain cost).

coupled power range   the allowable difference between the minimum and maximum allowed power.

coupling   connecting two fibers or the connector between two fibers.

cross-plug range   the difference between the measured lowest power and highest power for matings between multiple connectors and the same transceiver.

cross-talk   the leaking of or interference between signal in two nearby pixels, wires, or fibers.

cutoff wavelength   the wavelength for which the normalized power becomes linear to within 0.1 dB.

daisy chain   a bus wiring scheme where devices are connected to each other in sequence (device A is wired to device B is wired to device C, etc.), like a chain of daisies (see topology, bus).

dark current   the current measured from a detector when no signal is present.

data dependent jitter   see jitter.

data rate   short for data transfer rate, the speed devices transmit digital information. Units include bits per second, but are more likely to be in the range of megabits per second (Mbit/s), or even gigabits per second (Gbit/s) in the field of fiber optics.

decoding   see encoding/decoding.

detectivity   the reciprocal of the noise equivalent power (NEP). This can be a more intuitive figure of merit, because it is larger for more sensitive detectors.

deterministic jitter (DJ)   see jitter.

development expense   the total development investment in people, test equipment, prototypes, etc.

device   any machine or component that attaches to a computer.

dielectric mirror   a multilayer mirror that is an alternative to a distributed Bragg reflector (DBR) in semiconductor laser manufacture.

diffraction   when a wave passes through an edge or an opening, secondary wave patterns are formed which interfere with the primary wave. This can be used to create diffraction gratings, which work similarly to prisms.

directional   division multiplexing. See multiplexer.

dispersion   the distortion of a pulse due to different propagation speeds. Can be chromatic, which is caused by the wavelength dependence of the index of refraction, or intermodal, which is caused by the different paths traveled by the different modes.

distortion   the change in a signal's waveform shape.

distributed Bragg reflector (DBR)   a mirror used in the manufacture of semiconductor lasers and photodetectors. It is made from multiple layers of semiconductors that have a band gap in the wavelength of interest.

distribution panel   a central panel from which a signal is routed to points of use. See patch panel.

divergence   the bending of light rays away from each other, for example the spreading of a laser beam with increased distance.

doping   adding an impurity to a semiconductor.

dual-in-line pin package, butterfly package   a cavity package for a semiconductor laser that is wire bonded and then the plastic is molded around the body and leads of the package.

duplex   a communications line that lets you send and receive data at the same time.

duty cycle   the pulse duration times the pulse repetition frequency. In other words, the percentage of time an intermittent signal is on.

duty cycle distortion (DCD)   the ratio of the average pulse width of a bit to the mean of twice the unit interval (see unit interval).

dynamic range   the ratio of the largest detectable signal to the smallest detectable signal, such as the receiver saturation charge to the detection limit (also known as linear dynamic range).

electrooptic transducer   a device that converts an electric signal to an optical signal and vice versa. An example of this is the photocell and laser in a transceiver (see transducer).

emulation   the use of program to simulate another program or a piece of hardware.

encoding/decoding   encoding is the process of putting information into a digital format that can be transmitted using communications channels. Decoding is reversing the process at the end of transmission.

end node   a node that does not provide routing, only end user applications.

epitaxy   method of growing crystal layer on a substrate with an identical lattice, which maintains the continuous crystal structure. Styles include molecular beam epitaxy, vapor phase epitaxy, etc.

etalon   an optical device with two reflective mirrors facing each other to form a cavity.

etched mesa   a flat raised area on an electronic device, created during the photolithography process.

extinction ratio   the ratio of the power level of the logic "1" signal to the power of the logic "0" signal. It indicates how well available laser power is being converted to modulation or signal power.

eye diagram   an overlay of many transmitted responses on an oscilloscope, to determine the overall quality of the transmitter or receiver. The relative separation between the two logic levels is seen by the opening of the eye. Rise and fall times can be measured off the diagram. Jitter can be determined by constructing a histogram of the crossing point.

fabric, switched network a FCS network in which all of the station management functions are controlled at the switching point, rather than by each node. This approach removes the need for complex switching algorithms at each node. The telephone system, where the dialer supplies the phone number, can be used as an analogy.

Fabry-Perot resonance resonance occurs when waves constructively interfere. Fabry-Perot resonance occurs in a cavity surrounded by mirrors. Semiconductor lasers, such as VCSELs, take advantage of Fabry-Perot resonance to stimulate emission in their active region.

facet passivation growing a thin oxide layer over the semiconductor (for datacom the EELD) facet, to limit environmental exposure and natural oxidation.

Fermi level the maximum energy of the electrons in a solid, which determines the availability of free electrons. If the Fermi level is in the conduction (top) band, the material is a conductor (metal). If the Fermi level is in the valence (lower) band, the material is an insulator. If the Fermi level is between the conduction and the valence band, the material is a semiconductor. Ferrule- a cylindrical tube containing the fiber end that fits within high tolerance into the flange of the transceiver port.

fiber optic link the transmitter, receiver, and fiber-optic cable used to transmit data.

fiber ribbon multifiber cables and connectors.

Fibre Channel Connection (FICON) an I/O interface standard that mainframe computers use to connect to storage devices. This standard is eight times faster than the previous fiber optics standard, ESCON, due to a combination of new architecture and faster link rates (see architecture).

field installable connectors optical connectors that can be installed on-site at a customer location, as opposed to factory installed connectors that can only be attached at an authorized manufacturing center.

flicker noise noise with a $1/f$ spectrum, which occurs when materials are inhomogeneous.

flip chip mounting a semiconductor substrate in which all of the terminals are grown on one side of the substrate. It is then flipped over for bonding onto a matching substrate.

flip flop   a device which has two output states, and is switched by means of an external signal.

flow shop   a manufacturing area that makes only one product.

footprint   the amount of desk or floor space used by a component.

frame

1. In the SONET transmission format, a 125 microsecond frame contains 6480 bit periods, or 810 octets (bytes), that contain layers of information. These layers include the overhead, or instructions between computers, and the voice or audio channel. SONET can be used to support telephony.
2. A technique used by Web pages to divide the screen into multiple windows.

frame errors   errors from missing or corrupted frames, as defined by the ANSI Fibre Channel Standard.

frequency agile   capable of being easily adjusted over a range of operating frequencies.

frequency chirping   inducing either an up-chirp or down-chirp in an optical signal (see up-chirp and down-chirp).

frozen process   a process where, if there are major changes of equipment, process parameters, agents or even parts, a partial or total requalification of the process must be performed.

fusion splice   see splice.

gain   amplification. In a photo detector, the number of electron-hole pairs generated per incident photon.

GGP fiber   a specialty optical fiber, proprietary to 3M Corporation, which is more resistant to mechanical fractures when bent; used in optical connectors such as the VF-45.

Gigabit Ethernet (GBE)   a standard for high speed Ethernet that can be used in backbone environments to interconnect multiple lower speed internets (see backbone network).

Gigalink card   a laser based transceiver card that runs at approximately 1 Gbit/s and works as a transponder.

graded-index fiber   a fiber with a refractive index that varies with radial distance from the center.

gross profit   revenue minus cost of goods sold (COGS).

group velocity the transmission velocity of a wave packet, which is made of many photons with different frequencies and phase velocities.

hermetic seal a seal that air and fluids cannot pass through.

heterostructure a semiconductor layered structure, with lattice matched crystals grown over each other.

Hill gratings the first types of in-fiber Bragg diffraction gratings, named after the researcher who discovered them.

homologation confirming that a product follows the rules of each country in which it is used.

hubbed ring optical network architecture in which all data traffic flows through a single common location or hub (see topology).

hybrid integration a technique of developing smart pixels where optical devices are grown separately from the silicon electronic circuitry, and then are bonded together (see smart pixels).

image halftoning an image compression technique whereby a continuous-tone, gray-scale image is printed or displayed using only binary-valued pixels.

Infiniband an architecture standard for a high speed link between servers and network devices. It will initially run at 0.3 GB/s, but eventually scale as high as 6.0 GB/s. It is expected to replace peripheral component interconnect (PCI) (see architecture).

intelligent optical network optical network that also controls higher level switching and routing functions above the physical layer.

intermodal dispersion see dispersion.

intersymbol interference the distortion by a limited bandwidth medium on a sequence of symbols which causes adjacent symbols to interfere with each other.

intrinsic without impurities.

jitter The error in ideal timing of a threshold crossing event. The CCITT defines jitter as short-term variations of the significant instants (rising or falling edges) of a digital signal from their ideal position in time. Jitter can be both deterministic and random. Low frequency jitter can be tracked by the clock recovery circuit, and does not directly affect the timing allocations within a bit cell. Other jitter will affect the timing. Data dependent jitter, a type of deterministic jitter (DJ) includes intersymbol interference (ISI). DJ may also include DCD, sinusoidal jitter, and other non Gaussian jitter. Random jitter (RJ) can be defined as the peak to peak value of the bit error rate (BER) of 10^{-12}, or approximately 14 times the standard deviation of the Gaussian jitter distribution (see intersymbol interference, duty cycle distortion, bit error rate).

job shop   a manufacturing area that deals with a variety of products, each treated as a custom product.

jumper cable   an optical cable that provides a physical attachment between two devices or between a device and a distribution panel. In this way they are different from trunk cables.

junction capacitance   the capacitance formed at the pn junction of a photodiode.

lambertian   scattering that obeys Lambert's cosine law. The flux per unit solid angle leaving a surface in any direction is proportional to the cosine of the angle between that direction and the normal to the surface. Matt (not shiny) surfaces tend to be lambertian scatterers.

latency   the delay in time between sending a signal from one end of connection to the receipt of it at the other end.

lattice-matched   when growing a crystal layer on a substrate, the junction is lattice-matched if new material's crystal structure fits with the substrate's structure.

launch reference cable   a known good test cable used in loss testing.

layer   a program that interacts only with the programs around it. When a communications program is designed in layers, such as OSI, each layer takes care of a specific function, all of which have to occur in a certain order for communication to work (see layered architecture).

layered architecture   a modular way of designing computer hardware or software to allow for changes in one layer not to affect the others.

legacy product   a product which an organization has already invested in, or has currently installed. New technology must be compatible with legacy products.

line state   a continuous stream of a certain symbol(s) sent by the transmitter that, upon receipt of by another station, uniquely identifies the state of the communication line. For example, Q for quiet and H for halt.

linearity range   the range of incident radiant flux over which the signal output is a linear function of the input.

link   the fiber optic connection between two stations, including the transmitter, receiver, and cable, as well as any other items in the system, such as repeaters.

link budget   range of acceptable link losses.

link-level errors   errors detected at lower level of granularity than frames, as defined by the ANSI Fibre Channel Standard.

link loss analysis   a calculation of all the losses (attenuation) on a link.

liquid crystals, nematic, smectic, and cholesteric   a material which has some crystalline properties and some liquid properties. In optics these materials usually have elongated molecules that are rod shaped. If they are oriented randomly they have different optical properties than when they are aligned. Nematic liquid crystals tend to have the rods oriented parallel but their positions are random. Cholesteric liquid crystals are a subcategory of nematic, in which the molecular orientation undergoes a helical rotation about the central axis. Smectic liquid crystals have a permanent dipole moment that can be switched by an externally applied electric field. This gives them many photonics applications, such as smart pixel arrays.

local area network (LAN)   a network limited to about 1 km radius.

lock-in amplifier   a device used to limit noise by encoding input data with a known modulation, made by chopping the signal. The amplifier "knows" from this reference signal when the signal to be detected is on and when it is off. This allows the detector to low pass filter the data, which narrows the bandwidth of the detector, making it more precise (see chopper).

long wavelength   approximately 1300 nm (1270–1320) or 1550 nm.

loopback testing   looping a signal back across a section of the network to see if it works properly. If a transceiver passes a unit loopback test, but fails a network loopback test, the problem is in the cables, not in the transceiver.

loss   attenuation of optical signal.

machine capability   the statistical safety for processes performed by a tool or machine.

manufacturing cost the cost to manage procurement of parts, inventory, system assembly and test, and shipping costs.

margin the amount of loss, beyond the link budget amount, that can be tolerated in a link.

master/slave an architecture where one device (the master) controls other devices (the slaves).

mechanical splice see splice.

meshed rings network topology in which any node may be connected to any other node (see topology).

metrology the science of measurement.

metropolitan area network (MAN) interconnected LANs with a radius of less than 80 km (50 miles).

mini-zip a type of zipcord fiber cable with a smaller outer diameter than a standard zipcord.

mode mixing the changing of the modal power distribution following a splice.

mode partition noise within a laser diode, the power distribution between different longitudinal modes will vary between pulses. Each mode is delayed by a different amount due to the chromatic dispersion and group velocity dispersion in the fiber, which causes pulse distortion (see distortion, dispersion, and group velocity).

mode scrambled launch a type of optical coupling into a multimode fiber or waveguide that attempts to uniformly excite all modes in the target waveguide or fiber.

modulate/demodulate modulation is when one wave (the carrier) is changed by another wave (the signal). Demodulation is restoring the initial wave.

monolithic arrays a technique of simultaneous fabrication of electronic and optical circuits on the same substrate, to produce high speed smart pixels (see smart pixels).

Monte Carlo simulation any type of statistical simulation which accounts for the probability of various events. This technique is useful when dealing with non-Gaussian probability distributions. An example is the probability of absorption and scattering of photons traveling through a medium.

multimode fiber   a type of fiber in which light can travel in several independent paths.

multiplexer   a device that combines several signals over the same line. Wavelength division multiplexing (WDM) sends several signals at different wavelengths. Directional division multiplexing (DDM) combines the laser diode and photodiode using a coupler. Time compression multiplexing is similar to DDM, when a diode is in the transmission mode and the other in the receiving mode ("ping-pong transmission"). Space division multiplexing (SDM) requires two fibers—one for upstream transmission and one for downstream transmission. Time-division multiplexing (TDM) transmits more than one signal at the same time by varying the pulse duration, pulse amplitude, pulse position, and pulse code, to create a composite pulse train.

narrowcast (NC)   to direct a program to a specific, well defined audience.

nematic liquid crystals   see liquid crystals.

neural networks   a system of programs and data structures that simulates the brain. They use a large number of simple processors in parallel, each with local memory. The neural network is "trained" by feeding it data and rules about relationships between the data.

noise equivalent power (NEP)   the flux in watts necessary to give an output signal equal to the root mean square noise output of a detector.

noise floor   the amount of self-generated noise by a device.

numerical aperture (NA)   The NA defines the light gathering ability of a fiber, or an optical system. The numerical aperture is equal to the sine of the maximum acceptance angle of a fiber.

Nyquist frequency   the highest frequency that can be reproduced when a signal is digitized at a given sample rate. In theory, the Nyquist frequency is half of the sampling rate.

octet   eight bits in a row. Also known as a byte.

ongoing inventory management cost   the cost to maintain and manage an inventory, as compared with procuring parts for inventory stocking.

open rings   network topology in which a ring has been opened at one point and turned into a linear network with add/drop of channels at any point (see topology).

optical bus   a facility for transferring data between several fiber connected devices located between two end points, when only one device can transmit at a time.

optical bypass   an optical switch that diverts traffic around a given location.

optical interface   where the optical fiber meets the optical transceiver.

optical inter-networking   allows IP switching layer to operate at the same line rate as a DWDM network, typically using an OC-48c connection.

optical power   the time rate of flow of radiant energy of a signal, expressed in watts.

optical seam   part of a network that does not allow passthrough of traffic meant for other nodes.

overshoot   waveform excursions above the normal level.

packet switching   protocols where data is encoded into packets, which travel independently to a destination where they are decoded.

paradigm   a model, example or pattern.

parallel optical links   links that transform parallel electrical bit streams directly into parallel optical bit streams. Possible mechanisms include transceivers made from VCSELs and array detectors sending data through fiber ribbon (see VCSEL and fiber ribbon).

parts cost   the sum of all bill of material components used in a design.

passivation layer   to coat a semiconductor with an oxide layer, to reduce contamination by making the surface less reactive.

patch panel   a hardware unit that is used as a switch board, to connect within a LAN and to outside for connection to the Internet or a WAN.

phase-locked loop   a circuit containing an oscillator whose output phase locks onto and tracks the phase of a reference signal. The circuit detects any phase difference between the two signals and generates a correction voltage that is applied to the oscillator to adjust its phase. This circuit can be used to generate and multiply the clock signal (see clock generation/multiplication).

phase velocity   the speed of a wave, as determined by a surface of constant phase.

phasor formalism   a polar method of displaying complex (real and imaginary) quantities.

photoconductive   the mode of a reverse biased detector. This reduces the capacitance of the detector, and thus increases the speed of response of the diode. It is the preferred mode for pulsed signals.

photoconductor   a non-junction type semiconductor detector where incident photons produce free charge carriers, which change the electrical conductivity of the material. Lead sulfide and lead selenide are examples of this type of detector, as well as most MSMs.

photolithography   a process for imprinting a circuit on a semiconductor by photographing the image onto a photosensitive substrate, and etching away the background.

photomicrograph   a photograph of an object that is magnified more than 10 times its size.

photonics   the study of photon devices and systems.

photoresist implant mask   used in photolithography to cover parts of a photosensitive medium when imprinting a circuit on a semiconductor.

photovoltaic   the mode of a junction unbiased detector. Since $1/f$ noise increases with bias, this type of operation has better NEP at low frequencies.

pigtail   a short fiber permanently fixed on a component, used for connecting the component to the fiber optic system.

PIN (or p-i-n) photodiode   a diode made by sandwiching p-type (doped with impurities so the majority of carriers are holes), intrinsic (undoped), and n-type (doped with impurities so the majority of carriers are electrons) semiconductor layers. Photons absorbed in the intrinsic region create electron-hole pairs that are then separated by an electric field, thus generating an electric current in a load circuit.

plenum-rated cable   cable which can be used in a duct work system (plenum) which has smoke retardant properties.

plug repeatability   the variation in coupled power for multiple connections between the same components.

point-to-point transmission   carrying a signal between two endpoints without branching to other points.

polarized connector   a connector that can only plug in one position, so that it is aligned properly.

polarized light   light whose waves vibrate along a single plane, rather than randomly. There are several polarization modes such as TE, where the electric field is in the direction of propagation, and TM, where the magnetic field is in the direction of propagation.

preamplifier a low-noise amplifier designed to be located very close to the source of weak signals. Often the first stage of amplification.

private mode when a port or repeater receives only packets addressed to the attached node. Also known as normal mode.

process window a defined variation of process parameters that characterize equipment and production used in series production.

processor the part of a computer that interprets and executes instructions. Is sometimes used to mean microprocessor or central processing unit (CPU), depending on context.

product cost apportionment (PCA) or supply chain costs shipping and distribution expenses for raw material and assemblies from suppliers and delivery of some final product to the end customer.

promiscuous mode when a port or repeater passes on all packets, not only those addressed to the attached node.

protocol the procedure used to control the orderly exchange of information between stations on a data link, network, or system. There are several standards in code set, such as ASCII, transmission mode, asynchronous or synchronous, and non-data exchanges, such as contract, control, failure detection, etc.

pump laser diode a short wavelength (typically 900 nm range) laser used to provide a pump input or gain in an optical fiber amplifier.

quantum confined Stark effect a mechanism for changing the optical absorption of a quantum well by applying an electric field. Because of this effect quantum wells are used in optical modulators.

quantum efficiency (QE) the ratio of the number of basic signal elements produced by detector (usually photoelectrons) to the number of incident photons.

quantum well a heterostructure with sufficiently thin layers that quantum affects begin to affect the movement of electrons. This can increase the strength of electro-optical interactions by confining the carriers to small regions.

raised floor a floor made of panels which can be removed for easy access to the wiring and plumbing below.

Raman scattering   scattering off phonons (quanta of vibration). A special case is Brillouin scattering, which is when an acoustic phonon (sound wave) is involved.

random jitter (RJ)   see jitter.

refractive index   the ratio of the speed of light in a vacuum to the speed of light in a material at a given wavelength.

regeneration   insures that there is sufficient optical power for a signal to reach its destination.

repeater   a device placed in a data link to amplify and reshape the signal in mid-transmission, which increases the distance it can travel.

reshaping   removes pulse distortion caused by dispersion.

resonant cavity photodetectors   a photodetector made by placing a photodiode into a Fabry-Perot (FP) cavity to enhance the signal magnitude.

response time   the time it takes a detector's output to rise when subjected to a constant signal.

responsivity   the ratio of the detector output to the radiation input. It is usually expressed as a function of wavelength.

retiming   restores a timing reference signal to remove jitter and improve clock/data recovery.

ring topology   (see topology).

ringing   waveform overshoot and oscillations (see overshoot).

roadmap   a long range projection for the future of a type of product, for example "The National Technology Roadmap for Semiconductors."

run length   the number of consecutive identical bits, such as the number of 1's or 0's in a row, in the transmitted signal. The pattern 010111100 has a run length of four.

running disparity   the difference between the number of 1's and 0's in a character, which is often tracked as a special parameter in network management software.

saturation   when the detector begins to form less signal output for the same increase of input flux.

scalability   the ability to add power and capability to an existing system without significant expense or overhead.

Schottky-barrier photodiodes   a variation on the PIN photodiode where the top layer of semiconductor material has been eliminated in favor of a reverse biased, metal-semiconductor-metal (MSM) contact. This results in faster operation, but lower signal. The advantage of this approach is improved quantum efficiency, because there is no recombination of carriers in the surface layer before they can diffuse to either the ohmic contacts or the depletion region (see PIN photodiode).

scintillation   rapid changes in the irradiance of a laser beam.

serializer/deserializer   a serializer is a device that converts parallel digital information into serial. A deserializer converts it back.

server   a central computer where data is deposited, and can be accessed over the network by other computers, known as clients.

serviceability   the ease in which a product can be serviced and inspected.

short wavelength   780–850 nm.

Shot noise   noise made by the random variations in the number and speed of the electrons from an emitter.

shunt resistance   the resistance of a silicon photodiode when not biased.

signal-to-noise ratio   the ratio of the detector signal to the background noise.

simplex   a one-way communications line, which cannot both send and receive data.

single-mode fiber   fiber where the light can only propagate through one path.

skew   the tendency for parallel signals to reach an interface at different times. The skew for a copper cable is 2-nanosecond bit periods over 20 m, while for fiber ribbons it is under 10 picoseconds/m.

slope efficiency   the differential quantum efficiency of the laser combined with the losses of the optical coupling.

smart pixel array   an array of optical devices (either detectors, modulators, or transmitters) which are directly connected to logic circuits. By integrating both electronic processing and individual optical devices on a common chip, one may take advantage of the complexity of the electronic processing circuits and the speed of the optical devices.

smectic liquid crystals   see liquid crystals.

space division multiplexing   see multiplexer.

splice bushing   see bulkhead splice.

splice loss   the loss of optical signal due to splicing cables.

splices   to join together two pieces at their ends to form a single one. When applying to optical cable, to form a permanent joint between two cables, or a cable and a port. There are two basic types of splice, mechanical spices and fusion splices. Mechanical splices place the two fiber ends in a receptacle that holds them close together, usually with epoxy. Fusion splices align the fibers and then heat them sufficiently to fuse the two ends together.

spoofing   sending acknowledgements of data transfer before data is actually received at its destination; usually done to artificially reduce latency in a network.

spun fiber   a manufacturing process in which the fiber is rotated during the drawing process in an effort to remove polarization dependence.

star topology   see topology.

Stark effect   the splitting of spectral lines due to an incident electric field. The quantum confined Stark effect is a special case of that. It uses an electric field to wavelength modulate sensitivity of quantum well detectors (see quantum confined Stark effect).

statistical process control   the use of statistical techniques to analyze, monitor, and control a process. Quality is often monitored in this way.

storage area network (SAN)   a high speed network, or section of an enterprise network, of storage devices, for access by local area networks (LAN) and wide area networks (WAN).

strain relief   a design feature that relieves the pressure on a connector, which could otherwise cause it to crack or unplug.

striping   simultaneously allowing data transfer through multiple ports.

Strowger switch   developed in the late 1800s, this early telecommunications switch replaced human switchboard operators by automatically making phone connections using electrical signals on the phone line.

substrate   the base layer of support material on which crystals are grown. Products grown on substrates include semiconductor detectors and integrated circuits. A silicon wafer is an example of a substrate.

supply chain costs   see product cost apportionment (PCA).

surface wave filter   a surface acoustic wave (SAW) phase filter often used to generate and multiply the clock signal (see clock generation/multiplication).

switched base mode   operating mode of the IBM 2029 Fiber Saver DWDM device, in which unprotected channels are passed through a dual fiber optical switch that protects availability in the case of a fiber cut only, not in the case of equipment failure.

switched network   see fabric.

synchronous   a system is synchronous if it can send and receive data at the same time, using a common timing signal. Since there is regular time interval between transmissions, there is no need for a start or stop bit on the message.

thermal noise   noise cause by randomness in carrier generation and recombination due to thermal excitation in a conductor; it results in fluctuations in the detector's internal resistance, or in any resistance in series with the detector. Also known as Johnson or Nyquist noise.

time   compression multiplexing. See multiplexer.

time-division multiplexing   see multiplexer.

timing reference, trigger   signals used by the oscilloscope to start the waveform sweep.

TO can   metal packaging in the form of transistor outline for semiconductor lasers.

token   a frame with control information, which grants a network device the right to transmit.

topology   the physical layout of a network. The three most common topologies are bus, star, and ring. In a bus topology, all devices are connected to a central cable (see optical bus and backbone). In a ring topology, the terminals are connected serially point to point in an unbroken circle. In a star topology, all devices are connected to a central hub, which copies the signal to all the devices.

transceiver   a package combining a transmitter and a receiver.

transducer   a device that converts energy from one form to another. See electrooptic transducer.

transimpedance amplifier   an operation amplifier and variable feedback resistance connected between the input and the output of the amplifier. They are used to amplify low photodiode signals, and provide high dynamic range, good sensitivity, and bandwidth.

transponder   a receiver/transmitter which can reply to an incoming signal. Passive transponders allow devices to identify objects, such as credit card magnetic strips. Active transponders can change their output signal, such as radio transmitter receivers. Satellite systems use transponders to uplink signal from the earth, amplify it, convert it to a different frequency, and return it to the earth.

trigger   see timing reference.

troubleshooting   a systematic method to find the reason for a problem.

trunk fiber   a fiber between two switching centers or distribution points.

two beam holographic exposure   a technique used to fabricate gratings using an expanded laser beam that is divided by a beamsplitter and then recombined, creating an interference pattern. This pattern is transferred photolithographically onto the surface of a semiconductor substrate.

ultrasonic   sound waves at a frequency too high for humans to hear. These waves can be used to excite metals used in wire bonding, or to clean items before placing them in vacuum.

undershoot   waveform excursions below the normal level.

unit interval (UI)   the shortest nominal time between signal transition. The reciprocal of baud, it has units of seconds.

up-chirp   a linear increase in frequency over time.

vertical cavity surface-emitting lasers (VCSEL)   a semiconductor laser made from a bottom DBR, an active region, and a top DBR.

virtual tributary   a SONET format with lower bandwidth requirements, to allow services like a DS1 or T1 signal to be carried on a SONET path without remultiplexing the voice channels.

wafer   a flat round piece of silicon that is used as a substrate on which to manufacture integrated circuits.

wavelength-division multiplexing   see multiplexer.

wavelength-routed networks   a network with several routers which choose the light path of the signal in the network by its wavelength. This architecture is being developed for all-optical networking, and has WDM applications.

wide area network (WAN)   a network that is physically larger than a LAN, with more users.

wire bond   connecting wires to devices.

word   four contiguous bytes.

zipcord   a type of optical fiber consisting of two unibody cables connected in the middle by a thin, flexible out coating.

Some Useful Dictionaries on the World Wide Web

- The One Look Dictionary, www.onelook.com.
- The Photonics Dictionary, www.laurin.com/DataCenter/Dictionary/CD/index.htm.
- Academic Press Dictionary of Science and Technology, www.harcourt. com/dictionary/browse/.

Appendix B | Acronyms and Initials

2R	retime/reshape
3R	retime/reshape/regenerate
AAL	ATM adaptation layer
AC	alternating current
ACK	acknowledgment
ACTS	advanced communication technologies and services
A/D	analog to digital
AFNOR	Assoc. Française de NORmalisation (ISO member)
AIP	American Institute of Physics
ANS	American National Standards
ANSI	American National Standards Institute
AOI	add or invert
AON	all-optical network
APC	auto power control
APD	avalanche photodiode
API	application programmer interface
APS	American Physical Society
AR	antireflective
ARP	address resolution protocol
ARPA	Advanced Research Project Agency
ASE	amplified spontaneous emission
ASIC	application specific integrated circuit
ASCI	Advanced Strategic Computing Initiative
ASCII	American Standard Code for Information Interchange (developed by ANSI)

ASTM	American Society for Test and Measurement
ATM	asynchronous transfer mode
ATMARP	ATM address resolution protocol
AUA	another useless acronym (from Preface to the Second Edition)
AWG	array waveguide grating
BD	solid bore inner diameter
BER	bit error rate
BGA	ball grid array
BGP	border gateway protocol
BH	buried heterostructure
BiCMOS	bi-junction transistor/complementary metal oxide semiconductor
BIP	bit interleaved parity
BIS	Bureau of Indian Standards
B_ISDN	broadband integrated services digital network
bit/s	bits per second
BLIP	background limited in performance
BLM	ball-limiting metallurgy
BLSR	bidirectional line switched ring
BNR	Bell Northern Research
BOL	beginning of lifetime
BPS	bytes per second
BSI	British Standards Institute
BSY	control code for busy, as defined by the ANSI Fibre Channel Standard
BTW	behind the wall
BUS	broadcast and unknown server
BWPSR	bi-directional path-switched rings
C4	controlled collapse chip connection
CAD	computer aided design
CAM	computer aided manufacturing
CATV	community antenna television (cable TV)
C-band	conventional wavelength band used in wavelength multiplexing (1520–1570 nm)
CBD	connector body dimension
CBGA	ceramic ball grid array
CBR	constant bit rate

CCD	charge coupled devices
CCITT	Consultative Committee for International Telephone and Telegraph (International Telecommunications Standard Body)—now ITU
CCW	channel control words
CD	compact disc
CDL	converged data link
CDMA	code-division multiple access
CDR	clock and data recovery
CDRH	U.S. Center for Devices and Radiological Health
CE mark	Communaute Europeenne mark
CECC	CENELEC Electronic Components Committee
CEL	California Eastern Laboratories
CENELEC	Comitéé Europééen de Normalisation ELECtrotechnique or "European committee for electrotechnical standardization"
CFR	Center for Devices and Radiological Health
CICSI	Building Premises Wiring Standard Body
CIP	carrier induced phase modulation
CLEC	competitive local exchange carrier (local telephone service)
CLEI	common language equipment identification (Bellcore)
CLI	common line interface
CLO	control link oscillator
CLP	cell loss priority bit
CML	current-mode logic
CMOS	complementary metal oxide semiconductor
CMU	Carnegie Mellon University
CNS	Chinese National Standards
COB	chip on board
COGS	cost of goods sold
COSE	Committee on Optical Science & Engineering
CPR	coupled power range
CPU	central processing unit
CRC	clock recovery circuit
CRC	cyclic redundancy check
CS	convergence sublayer
CSMA/CD	carrier sense multiple access/with collision detection
CSP	channeled substrate planar
CTE	coefficient of thermal expansion
CVD	chemical vapor deposition

CW	continuous wave
CWDM	coarse wavelength-division multiplexing (also known as wide-WDM)
CYTOP	a transparent fluorpolymer
DA	destination address
DARPA	Defense Advanced Research Projects Association (see also ARPA)
DAS	dual-attach station
DASD	direct access storage device
DBR	diffraction Bragg reflector
DBR	distributed Bragg reflector
DC	direct current
DCD	duty cycle distortion, also data carrier detect (modems)
DCF	dispersion compressing fiber
DCN	data communications network
DDM	directional division multiplexing
DDOA	dysprosium doped optical amplifier
DE	diethyl (as in diethylzinc-DEZn)
DEC	Digital Equipment Corporation
DEMUX	demultiplexer
DFB	distributed feedback
DH	double heterojunction/double heterostructure
DHHS	Department of Health Services in the IEC
DIN	Deutsches Institut für Normung, (the German Institute for Standardization) or Data InputE06.html
DIP	dual in-line package
DJ	derministic jitter
DLL	delay locked loop
DM	dimethyl (as in dimethylzinc-DMZn)
DMD	differential mode delay
DMM	digital multimeter
DP	data processing
DRAM	dynamic random access memory
DS	Dansk Standardiseringsrad
DSF	dispersion shifted fiber
DSL	digital subscriber line
DSO	digital sampling oscilloscope
DSP	digital signal processing
DTPC	design for total product cost

DUT	device under test
DVD	digital video disk
DWDM	dense wavelength division multiplexers
EBP	end of bad packet (delimiter for Infiniband data packets)
EC	European Community
ECF	echo frame
ECH	frames using the ECHO protocol
ECHO	European Commission Host Organization
ECL	emitter coupled logic
EDFA	erbium doped fiber amplifier
EELD	edge-emitting semiconductor laser diode (also ELED)
EFA	endface angle
EGP	end of good packet (delimiter for Infiniband data packets)
EH	hybrid mode where electric field is largest in transverse direction
EIA	environmental impact assessments
EIA	Electronics Industry Association
EIA standard	Electronics Industry Association standard
EIA/TIA	Electronics Industry Association/Telecommunications Industry Association
ELAN	emulate local area network
ELED	edge-emitting semiconductor laser diode (also EELD)
ELO	epitaxial liftoff
EMBH	etched-mesa buried heterostructure
EMC	electromagnetic compatibility
EMI	electromagnetic interference
ENIAC	Electronic Numeric Integrator and Computer
EOF	end of frame
EOL	end of lifetime
ESCON	Enterprise system (IBM) connection
ESD	electrostatic discharge
ESPRIT	European Strategic Program for Research in Information Technology
ETDM	electronic time division multiplexing
ETR	external timing reference
ETSI	European Telecommunication Standards Institute

FBG	fiber Bragg grating
FBS	fiber beam spot
FC	fibre channel
FC	frame control
FC connector	threaded fasteners
FC	flip chip, a manufacturing chip bonding procedure
FCA	Fibre Channel Association
FC-AL	Fibre Channel arbitrated loop
FCC	Federal Communications Commission
FCE	ferrule/core eccentricity
FCS	Fibre Channel Standard
FCSI	Fibre Channel Standard Initiative
FCV	Fibre Channel Converted/ FICON conversion vehicle
FD	ferrule diameter
FDA	Food and Drug Administration
FDDI	fiber distributed data interface
FDMA	frequency division multiple access
FET	field effect transistor
FF	ferrule float
FICON	Fibre Channel Connection or Fiber Connection
FIFO	first in, first out
FIG	fiber image guides
FIT	failures in time
FJ	Fiber-Jack
flops	floating point operation performed per second
FLP	fast link pulse
FO	fiber optic
FOCIS	fiber optic connector intermatability standard
FOTP	fiber optic test procedures
F_Port	fabric port, as definded by the ANSI Fibre Channel Standard
FP	Fabry-Perot
FPA	Fabry-Perot amplifier
FPGA	field programmable gate array
FQC	Fiber Quick Connect (feature in the IBM Global Services structured cable offering)
FRPE	flame-retardant polyethylene
FSAN	full service access network
FSR	free spectral range

FTC	fiber quick connect
FTS	fiber transport services
FTTC	fiber-to-the-curve
FTTH	fiber-to-the-home
FTTO	fiber-to-the-office
FWHM	full width at half maximum
FWM	four wave mixing

GBE or GbE	Gigabit Ethernet
GBIC	gigabit interface connector
GDPS	geographically dispersed Parallel Sysplex
GFC	generic flow control
GHz	gigahertz
GI	graded index
GLC	Gigalink Card
GOSS	State Committee of the Russian Federation for Standardization Metrology and Certification
GP	gross profit
GRIN	graded index
GRP	glass reinforced plastic
GVD	group velocity dispersion

HAN	home area network
HC	horizontal cross-connect
HE	hybrid mode where magnetic field is largest in transverse direction
HEC	header error correction
HFET	heterostructure field effect transistor
HIBITS	high bitrate ATM termination and switching (a part of the EC RACE program)
HIPPI	high performance parallel interface
HMC	hardware management console
HOCF	hard polymer clad (glass) fiber
HOEIC	hybrid optoelectronic integrated circuit
HPC	high performance computing
HPCF	hard polymer clad interface
HR	highly reflective
HSPN	High Speed Plastic Network Consortium

IB	InfiniBand
IBM	International Business Machines

IC	integrated circuit
IC	intermediate cross-connect
ICCC	International Conference on Computer Communications
ID	identification
IEC	International Electrotechnical Commission
IEEE	Institute of Electrical and Electronics Engineers
IETF	Internet Engineering Task Force
IFB	imaging fiber bundles
IHS	Information Handling Services
ILB	inner lead bond
ILEC	incumbent local exchange carrier (local telephone service)
ILMI	interim local management interface
InATMARP	inverse ATM address resolution protocol
I/O	input/output
IODC	Institute for Optical Data Communication
IP	internet protocol
IPI	intelligent physical protocol
IPT	integrated photonic transport (inband)
IR	infrared reflow
IrDA	Infrared Datacom Association
IRED	infrared emitting diode
ISC	intersystem channel gigabit links
ISI	intersymbol interference
ISO	International Organization for Standardization
ISP	internet service provider
IT	information technology
ITO	indium tin oxide
ITS	Institute for Telecommunication Sciences
ITU	International Telecommunications Union
IVD	inside vapor deposition
IXC	interexchange carrier (long distance telephone service)
JEDEC	Japanese based telecom/datacom standards
JIS	Japanese Industrial Standards
JSA	Japanese Standards Association
KGD	known good die

LAN	local area network
LANE	local area network emulation
L-band	long wavelength band used in wavelength multiplexing (1560–1610 nm)
LC	late counter
LC	"Lucent Connector"
LCOS	liquid crystal on silicon
LCT	link confidence test
LD	laser diode
LEAF	large effective area fiber
LEC	LAN emulation client
LECID	LEC (LAN emulation client) identifier
LED	light emitting diode
LES	LAN emulation server
LIA	Laser Institute of America
LIS	logical IP subnet
LLC	logic link control
LOH	line overhead
LOS	loss of signal
LP	linear polarized mode
LPE	liquid phase epitaxy
LSZH	low smoke zero halogen cable
LUNI	LANE user-network interface
LVD	low voltage directive
LVDS	low voltage differential signal
LX	long wavelength (1300 nm) transmitter
MAC	media access control
MAN	metropolitan area network
MBE	molecular beam epitaxy
MBGA	metal ball grid array
MC	main cross-connect
MCM	multichip module
MCP	mode conditioning patch cables
MCVD	modified chemical vapor deposition
MDF	main distribution facility
MDI	medium dependent interfaces
MEM	microelectromechanical
MESFET	metal semiconductor field effect transistor
METON	metropolitan optical network

MFD	modal field diameter
MIB	management information base
MIC	media-interface connector
MII	media-independent interface
MIMD	multiple instruction multiple data stream
MJS	methodology for jitter specification
MM	multimode
MMF	multimode fiber
MOCVD	metal organic chemical vapor deposition
MONET	multiple wavelength optical network
MOS	metal oxide semiconductor
MOSIS	MOS implementation system
MPE	maximum permissible exposure
MPLS	multi-protocol label switching
MPLmS	multi-protocol lambda switching
MPO	trade name for multifiber optical connector using a 12-fiber interface; also known as MPX
MPX	trade name for multifiber optical connector using a 12-fiber interface; also known as MPO
MQW	multiquantum well
MS	multi-standard
MSA	multi-source agreements
MSI	medium scale integration
MSM	metal-semiconductor-metal
MT	multifiber termination
MTP	multifiber terminated push-on connector
MT-RJ	multi-termination RJ-45 latch (a type of optical connector)
MT-RT	multifiber RJ-45 latched connector
MU	multi-termination unibody
MUX	multiplexer
MZ	Mach-Zehnder
NA	numerical aperture
NAND	not and (logic gate)
NAS	network attached storage
NBS	National Bureau of Standards (now NIST)
NC	narrowcast
NC&M	network control and management
NEBS	network equipment building system

NEC	National Electric Code
NEP	noise equivalent power
NEXT	near end crosstalk
NFPA	National Fire Protection Agency
NGI	next generation Internet
NGIO	next generation input/output
NIC	network interface card
NIF	neighbor information frame
NIST	National Institute of Standards & Technology
NIU	network interface units
NLOG	non-linear optical gate
NMOS	negative channel metal oxide semiconductor
NMS	network management system
NOR	not or (logic gate)
N_Port	nodal port, as defined by the ANSI Fibre Channel Standard
NPN	negative-positive-negative
NRC	National Research Council
NRZ	nonreturn to zero
NSPE	National Society of Professional Engineers
NTIA	National Telecommunication and Information Administration
NTT	Nippon Telegraph and Telephone Corporation
NUMA	non-uniform memory architecture
NZDSF	non-zero dispersion shifted fiber
OA	optical amplifiers
OADM	optical add-drop multiplexer
OAMP	operations, administration, maintenance, and provisioning (sometimes OAM&P)
OCI	optical channel interface
OCI	optical chip interconnect
OCLD	optical channel laser detector
OCM	optical channel manager
ODC	optical data center
ODSI	Optical Domain Service Interconnect coalition
OE	opto-electronic
O/E/O	optical/electrical/optical
OFB	ordered fiber bundles
OFC	optical fiber control system

OFN	optical fiber nonplenum/nonriser/nonconductive
OFNP	optical fiber nonconductive plenum listing
OFNR	optical fiber nonconductive riser listing
OFSTP	optical fiber system test procedure
OIDA	Optoelectron Industry Development Association
OIF	Optical Internetworking Forum
OIIC	optically interconnected integrated circuits
OLIVES	optical interconnections for VLSI and electronic systems (a part of the EC ESPRIT program)
OLS	optical label switching
OLTS	optical loss test set
OMA	optical modulation amplitude
OMB	office of management and budget
OMC	operational management committee
OMNET	optical micro-network
OMX	optical multiplexing modules
ON	Osterreisches Normungistitut (Austrian Standards Institute)
ONTC	Optical Network Technology Consortium
OPFET	optical field effect transistor
OSA	open system adapter
OSA	optical sub assembly, also called the coupling unit
OSA	Optical Society of America
OSHA	Occupational Safety & Health Administration
OSI	open systems interconnection
OSIG	optical signal
OSNR	optical signal to noise ratio
OSPF	open shortest path first
OTDM	optical time division multiplexer
OTDR	optical time domain reflectometer
OTF	optical transfer function
OTN	optical transport network
OUT	output
OUTB	average output
OVD	outside vapor deposition
OXBS	optical crossbar switches
OXC	optical cross connects
PA	preamble
PA	pointing angle

PANDA	polarization maintaining and absorption reducing fiber
PAROLI	parallel optical link
PBGA	plastic ball grid array
PBX	private branch exchange
PC	personal computer
PC	polycarbonate-a thermoplastic compound
PCA	product cost apportionment
PCB	printed circuit board
PCI	peripheral component interconnect
PCM	physical connection management
PCVD	plasma-assisted chemical vapor deposition
PD	photodiode
PDF	probability density function
PDG	polarization dependent gain
PDL	polarization dependent loss
PECL	postamplifier emitter coupled logic unit
PECVD	plasma-enhanced chemical vapor deposition
PF	perfluorinated
PGA	pin-grid arrays
PHY	physical layer of OSI model
PIN	positive-intrinsic-negative
PLCC	plastic leaded chip carrier
PLL	phase-lock loop
PLOU	physical layer overhead unit
PLS	primary link station
PMI	physical media-independent sublayer
PMD	physical medium-dependent sublayer
PMD	polarization mode dispersion
PMF	parameter management frame
PMF	polarization maintaining fiber
PMI	physical medium independent sublayer
PMMA	polymethylmethacrylate—a thermoplastic compound
PMT	photomultiplier tube
POF	plastic optical fiber
POI	parallel optical interconnects
POINT	polymer optical interconnect technology
PON	passive optical network
POP	post office protocol
POTS	plain old telephone system
PPL	phase locked loop

PPRC	peer to peer remote copy
PR	plug repeatability
PRBS	pseudo-random binary sequence
PSTN	packet switched telephone network
PT	payload type
PTT	Public telephone and telegraph
PVC	polyvinyl chloride (a plastic)
PVL	parallel vixel link
QCSE	quantum confined Stark effect
QE	quantum efficiency
QFP	quad flat pack
QOS	quality of service
QW	quantum well
RACE	Research in Advanced Communications for Europe
RAID	redundant array of inexpensive disks
RARP	reverse address resolution protocol
RC	resistor capacitor
RCDD	registered communication distribution designers
RCE	resonant-cavity enhanced
RCLED	resonant-cavity light emitting diode
RECAP	resonant-cavity enhanced photodetector
RF	radio frequency
RFI	radio frequency interference
RHEED	reflection high-energy electron diffraction
RIE	reactive ion etching
RIN	reflection induced intensity noise
RISC	reduced instruction set computing
RJ	random jitter
RJT	control code for reject, as defined by the ANSI Fibre Channel Standard
RMAC	repeater media access control
RMS	root mean square
ROSA	receiver optical subassembly
RSOH	regenerator section overhead
Rx	receiver
SA	source address
SA	Standards Association of Australia
SAGCM	separate absorption, grading, charge sheet and multiplication structure

SAM	separate absorption and multiplication layers
SAM	sub-assembly misalignment
SAN	storage area network
SAR	segmentation and reassembly sublayer
SAS	single attach station
SASO	Saudi Arabian Standards Organization
S-band	short wavelength band used in wavelength multiplexing (1450–1510 nm)
SBCCS	single byte command code set
SBCON	single byte command code sets connection architecture
SBS	source beam spot
SBS	stimulated Brillouin scattering
SC connector	subscriber connector (spring latch fasteners)
SC-DC	subscriber connector, dual connect ferrule
SCI	scalable coherent interface
SC-QC	subscriber connector, quad connect ferrule
SCSI	small computer system interface
SD	shroud dimension
SD	start delimiter
SDH	synchronous digital hierarchy
SDM	space division multiplexing
SDO	standards developing organizations
SEAL	simple and efficient adaptation layer
SEED	self electro-optic effect devices
SETI	search for extraterrestrial intelligence
SFF	small form factor
SFP	small form factor pluggable
SFS	Suomen Standardisoimislitto Informaatiopalvelu (Finland Standards Information)
SI	System International (International System of Units)
SIA	Semiconductor Industry Alliance
SIC	Standard Industrial Classification
SIF	status information frame
SIMD	single instruction-stream, multiple data-stream
SiOB	silicon optical bench (non-hermetic)
SIPAC	Siemens Packaging System
SIS	Standardiseringkommisiionen I Sverige (Swedish Standards Commission)
SJ	sinusoidal jitter

SL	superlattice
SLA	secure level agreement
SM	single mode
SMC	surface mount component
SMD	surface mount device
SMDS	switched megabit data services
SMF	single-mode fiber
SMP	shared memory processor
SMT	station management frames
SMT	surface mount technique
SMU	Sanwa multi-termination unibody
SNMP	simple network management protocol
SNR	signal-to-noise ratio
SNZ	Standards Association of New Zealand
SOF	start of frame
SOHO	small office/home office
SOJ	small outline packages with J leads
SONET	Synchronous Optical NETwork
SP	shelf processor
SPC	statistical process control
SPE	synchronous payload envelope
SPIBOC	standardized packaging and interconnect for inter and intra board optical connections (a part of the EC ESPRIT program)
SPIE	Society of Photooptic Instrumentation Engineers
SQNR	signal-to-quantum-noise ratio
SQW	single quantum well
SRAM	static random access memory
SRM	sub-rate multiplexing
SRS	stimulated Raman scattering
SSA	serial storage architecture
SSA	storage system architecture
S-SEED	symmetric self electro-optic effect devices
SSI	small scale integration
SSP	storage service provider
ST	connector–subscriber termination
STP	shielded twisted pair
STS	synchronous transport signal
SWDM	sparse WDM
SWF	surface wave filter
SX	short wavelength (850 nm) transmitter

TAB	tape access bonding
TAB	tape automated bonding
TAXI	transparent asynchronous transceiver/receiver interface
TBGA	Tape ball grid array
TC	transmission convergence sublayer
TCM	time compression multiplexing
TCP/IP	transmission control protocol/internet protocol
TDFAs	thulium doped fiber amplifiers (to be used in the wavelength range of 1450–1510 nm)
TDM	time division multiplexing
TDMA	time division multiple access
TE mode	transverse electric (electric field normal to direction of propagation)
TE	triethyl (as in triethylgallium—TEGa)
TERKS	trade name of a data base used by Telcordia (formerly Bell Labs) to track inventory and depreciation on telecommunications hardware
THC	through-hole component
THT	token holding timer
THz	tera hertz
TIA	Telecommunications Industry Association
TIA	trans-impedance amplifier (also TZA)
TM mode	transverse magnetic (magnetic field normal to direction of propogation)
TM	trimethyl (as in trymethylgallium-TMGa)
TO	transistor outline
TOSA	transmitter optical subassembly
TP	twisted pair
TRT	token rotation timer
TRTT	target token rotation time
TRX	transceiver
TTI	time to installation
TTL	transistor-transistor logic
TUG	tributary unit groups
TWA	traveling wave amplifier
Tx	transmitter
TZA	trans-impedance amplifier (also TIA)
UDP	user datagram protocol
UHV	ultra high vacuum

UI	unit interval
UL	Underwriters Laboratories
ULP	upper level protocol
UNI	user–network interface
USNC	United States National Committee
UTP	unshielded twisted pair
UV	ultraviolet
VAD	vapor axial deposition
VBR	variable bit rate
VC	virtual circuit
VCC	virtual channel connection
VCI	virtual channel identifier
VCO	voltage-controlled oscillator
VCSELs	vertical cavity surface emitting lasers
VDE	Verband Deutscher Electrotechniker (Association of German Electrical Engineers)
VDI	Vereins Deutscher Ingenierure (German Standards)
VF-45	trade name of a small form factor optical connector developed by 3M corporation with a RJ-45 latch
VLIW	very long instruction word
VLSI	very large-scale integration
VPI	virtual path identifier
VPN	virtual private network
VPR	vapor phase reflow
VT	virtual tributary
VTG	virtual tributary group
WAN	wide area network
WDM	wavelength-division multiplexing
WWSM	wide spectrum WDM
XDF	extended distance feature
XGP	Multi-Gigabit Ethernet pluggable
XPR	cross plug range
XRC	extended remote copy
ZBLAN	fluorozirconate

Some useful acronym searches on the World Wide Web

- The One Look Dictionary, www.onelook.com.
- Stammtisch Beau Fleuve Acronyms http://www.plexoft.com/SBF/ (best accessed through The One Look Dictionary).
- Acronym Finder http://www.acronymfinder.com/.

Appendix C | Measurement Conversion Tables

English-to-Metric Conversion Table

English unit	Multiplied by	Equals metric unit
Inches (in.)	2.54	Centimeters (cm)
Inches (in.)	25.4	Millimeters (mm)
Feet (ft)	0.505	Meters (m)
Miles (mi)	1.61	Kilometers (km)
Fahrenheit (F)	$(F - 32) \times 0.556$	Celsius (C)
Pounds (lb)	4.45	Newtons (N)

Metric-to-English Conversion Table

Metric unit	Multiplied by	Equals English unit
Centimeters (cm)	0.39	Inches (in.)
Millimeters (mm)	0.039	Inches (in.)
Meters (m)	3.28	Feet (ft)
Kilometers (km)	0.621	Miles (mi)
Celsius (C)	$(°C \times 1.8) + 32$	Fahrenheit (F)
Newtons (N)	0.225	Pounds (lb)

Absolute Temperature Conversion

Kelvin (K) = Celsius +273.15
Celsius = Kelvin −273.15

Area Conversion

1 square meter = 10.76 square feet = 1550 square centimeters
1 square kilometer = 0.3861 square miles

Metric Prefixes

Yotta = 10^{24}
Zetta = 10^{21}
Exa = 10^{18}
Peta = 10^{15}
Tera = 10^{12}
Giga = 10^{9}
Mega = 10^{6}
Kilo = 10^{3}
Hecto = 10^{2}
Deca = 10^{1}
Deci = 10^{1}
Centi = 10^{2}
Milli = 10^{-3}
Micro = 10^{-6}
Nano = 10^{9}
Pico = 10^{12}
Femto = 10^{15}
Atto = 10^{18}
Zepto = 10^{21}
Yotto = 10^{24}

Speed of light = c = 2.99792458×10^{8} m/s
Boltzmann constant = k = 1.3801×10^{23} J/K = 8.620×10^{5} eV/K
Planck's constant = h = 6.6262×10^{34} J/S
Stephan–Boltzmann constant = σ = 5.6697×10^{8} W/m^{2}/K^{4}
Charge of an electron = 1.6×10^{19} C
Permittivity of free space = 8.849×10 F/M
Permeability of free space = 1.257×10^{6} H/m
Impedance of free space = 120π ohms = 377 ohms
Electron Volt = 1.602×10^{19} J

Index